FARADAY

G000099589

FARADAY

Geoffrey Cantor
David Gooding
and
Frank A. J. L. James

MACMILLAN

© Geoffrey Cantor, David Gooding and Frank A. J. L. James 1991

All rights reserved. No reproduction, copy or transmission
of this publication may be made without written permission.

No paragraph of this publication may be reproduced, copied or
transmitted save with written permission or in accordance with
the provisions of the Copyright, Designs and Patents Act 1988,
or under the terms of any licence permitting limited copying
issued by the Copyright Licensing Agency, 33–4 Alfred Place,
London WC1E 7DP.

Any person who does any unauthorised act in relation to
this publication may be liable to criminal prosecution and
civil claims for damages.

First published 1991

Published by
MACMILLAN EDUCATION LTD
Houndmills, Basingstoke, Hampshire RG21 2XS
and London
Companies and representatives
throughout the world

Edited and typeset by Povey/Edmondson
Okehampton and Rochdale, England

Printed in Hong Kong

British Library Cataloguing in Publication Data
Cantor, Geoffrey *1943–*
Faraday
1. Physics. Faraday, Michael, 1791–1867
I. Title II. Gooding, David *1947–* III. James, Frank A.
J. L. *1955–*
530.092
ISBN 0–333–54291–6

Contents

List of Figures

List of Plates

Preface

Michael Faraday (1791–1867) made important experimental discoveries, such as the principles behind the electric motor, transformer and dynamo, and he introduced novel and insightful scientific theories, such as field theory. He was also an effective communicator of science. The Royal Institution Christmas Lectures, which are now seen on television by millions of people each year, were founded by Faraday. This short biography follows Faraday's lead since it aims to show, in non-technical language, how one major scientist lived and worked. It is a contribution to the public understanding of science and also to the bicentenary celebrations of Faraday's birth.

The authors and publishers wish to thank the Director of the Royal Institution for permission to reproduce figures from Faraday's notebooks and plates 1, 3, 4, 6 and 7; the Hunterian Art Gallery, Glasgow, for premission to publish plate 2; the President and Council of the Royal Society for permission to publish plate 5; Dr Philip Embleton for plate 8; and the Educational Services Unit of the University of Bath for producing the illustrations from Faraday's laboratory *Diary* and other drawings in Chapters 4 and 5.

Royal Institution

Geoffrey Cantor
David Gooding
Frank A.J.L. James

Key References

The main works we have cited are:

Faraday, *Experimental Researches in Electricity* (3 vols., London, 1839–55: facsimile reprint, New York: Dover, 1965). Cited as *Researches.*

Faraday, *Experimental Researches in Chemistry and Physics* (London, 1859). This contains over 50 of Faraday's papers on miscellaneous subjects and is cited as *ERCP.*

T. Martin (ed), *Faraday's Diary. Being the Various Philosophical Notes of Experimental Investigation made by Michael Faraday, DCL, FRS, during the Years 1820–1862 and Bequeathed by him to the Royal Institution of Great Britain* (7 vols. + index, London, 1932–6). Cited as *Diary.*

L. P. Williams, R. FitzGerald and O. Stallybrass, *The Selected Correspondence of Michael Faraday* (2 vols, Cambridge: Cambridge University Press, 1971). Cited as *Selected Correspondence.*

Frank A. J. L. James, *The Correspondence of Michael Faraday. Volume 1, 1811-December 1831, Letters 1–524* (Stevenage: Peter Peregrinus, 1991). Cited as *Correspondence.*

G. W. A. Kahlbaum and F.V. Darbishire, eds, *The Letters of Faraday and Schoenbein, 1836–62* (Basle and London, 1899). Cited as *Schoenbein.*

H. Bence Jones, *The Life and Letters of Faraday* (2 vols, London, 1870). Cited as Jones.

1 Introduction

Meet Michael Faraday. Plate 1 (pages 2–3) shows him delivering a lecture in the packed theatre of the Royal Institution in London. The date is the winter of 1855–6 and Faraday is now 64 years old, having been in the forefront of British science for a quarter of a century. Tall, distinguished and authoritative, he commands the attention of his audience (except for one easily-distracted boy) as he instructs them about the properties of metals (hence the charts headed 'silver' and 'gold' to which he is pointing). He is known to his audience as one of the foremost experimentalists and 'men of science'[1] in Britain. He has not only made many startling discoveries, particularly in the areas of chemistry, electricity and magnetism, but also possesses an unrivalled public reputation as a lecturer and as a person of high integrity. We see him here with left arm outstretched, caught in a moment of time.

How did this man, the son of a poor blacksmith, come to be lecturing in the Royal Institution? What was his background? Where did he receive his education in science? And how did he make the many discoveries that are associated with his name? These and similar questions will be discussed throughout this book, but for the present we will briefly outline his background.

Faraday was born on 22 September 1791 in London, in the area now known as the Elephant and Castle, although the family soon moved to lodgings off Welbeck Street, in the West End. His parents had recently moved to London from Westmorland (now part of Cumbria) and his father worked in a smithy. The family had little money and young Faraday received only a rudimentary schooling in reading, writing and arithmetic. At the age of 14 he commenced a seven-year apprenticeship with a bookbinder. In Faraday's day – unlike today – few people studied science in school or university and even fewer pursued a scientific career. Yet by reading and attending lectures Faraday gained a grounding in science and, through hard work and deep commitment to the subject, became the eminent,

Plate 1 Alexander Blaikley's group portrait

GOLD

polished scientist we see in the picture. Faraday's transition from the son of a poor ironsmith to the doyen of British science will be the subject of the next chapter.

Faraday's personal history cannot be dissociated from his social and institutional context. In the portrait we see him standing in the lecture theatre of the Royal Institution which was founded at the end of the previous century to spread scientific knowledge and its technological innovations to London's populace. In 1813 Faraday was appointed, at the suggestion of Sir Humphry Davy, Chemical Assistant at the Royal Institution. He subsequently spent the rest of his life not only working there but also living in the same building. He and his wife, Sarah, had rooms above the lecture theatre, while the laboratory in which he performed his celebrated researches was in the basement of this building. The lecture theatre in the picture was therefore home territory for Faraday.

The people close to Faraday were also often associated with the Royal Institution: thus his brother and nephew had manufactured and installed the gas lighting shown illuminating the theatre, and his faithful laboratory assistant, Sergeant Charles Anderson, is painted standing just behind him. Sarah is sitting in the right foreground wearing a dark bonnet. Also present, but not in the picture, is the artist, Alexander Blaikley, who was both a distant relative of Faraday's and a member of the small Christian sect, called the Sandemanians, to which Michael and Sarah belonged.

As we shall see in Chapter 3, Faraday was also connected with other institutions, such as the Royal Society (where he presented his main researches to other scientists), the British Association, the Royal Military Academy (where he also lectured) and Trinity House (which is responsible for maintaining lighthouses round the coast of England and Wales). Although an active, respected member of the early-Victorian scientific community Faraday differed from his contemporaries in a number of significant ways. For example, he worked in relative isolation and refused to have any students. He also played a minor role in the organisation of British science and generally avoided becoming involved in its politics. Contemptuous of the British honours system, he even declined the Presidencies of both the Royal Society and the Royal Institution. He was idiosyncratic in other ways; for example, his scientific papers contained virtually no mathematics at a time when physics was becoming increasingly mathematised.

Look again at the picture, particularly the audience. There are a number of children present since this was one of the annual series of juvenile science lectures that Faraday established at the Royal Institution in the late 1820s. Today you can see his successors' lectures televised after Christmas from this same theatre. Faraday was not only noted for his annual series of science lectures for children, but he was also the foremost communicator of science in his generation. He firmly believed that scientists should not only talk shop among themselves but that, as an essential part of their social duty, they must also make their findings accessible and interesting to the wider pubic. He gave considerable thought to the presentation of his lectures which were profusely illustrated by experiments. Faraday's success in the role of communicator has led the Royal Society recently to found the Faraday Award which is bestowed, on the recommendation of the Committee on the Public Understanding of Science, on those scientists who contribute significantly to the public understanding of science.

A selection of Faraday's public is shown in the picture. There is a high proportion of women in the audience. Among the eminent scientists present are the astronomer Sir James South, the geologists Sir Roderick Murchison and Sir Charles Lyell, the chemist Lyon Playfair and the physicist John Tyndall. Tyndall was Professor of Natural Philosophy at the Royal Institution. Others in the portrait include Rev. John Barlow (Secretary) and Faraday's trusted assistant Charles Anderson. Also present are several members of Faraday's family including Sarah and her brother George who was an artist and art teacher. Look closely at the man in the first row directly in front of the lecture table. He is Albert, soon to be Prince Consort, who had received a good education in science in his native Germany. Albert greatly admired Faraday, attended a number of his lectures and invited him on many occasions to tutor members of the Royal family in science. Faraday delivered lectures not only to Albert but also to Queen Victoria and the young princes: to Albert's left is sitting the inattentive Prince of Wales (later Edward VII). Albert's personal enthusiasm for science contrasts strikingly with the general disinterest in science among subsequent Royalty.

There is another, more significant, contrast. Today, science is generally viewed as the province of white-coated experts, while the man or woman in the street is debarred from entering its sanctuary. Moreover, the contrast is often drawn between science and culture.

However, as Blaikley's picture shows, the rich, the famous and the fashionable flocked to Faraday's lectures. There were many other places where science lectures were regularly delivered, both in London and in the provinces, and articles by leading scientists appeared frequently in the Victorian periodical press. A much closer relationship existed between the scientist and the public or, to put the matter another way, science was far more closely integrated with the rest of culture in the early Victorian period than it is today.

Turn your attention to the table. Here are samples of metals and pieces of simple demonstration apparatus connected with his lecture which dealt with the properties of common metals. Had we come on another day we might have found the bench covered with electrical apparatus, perhaps illustrating how the current in a wire affects a neighbouring magnetic needle or how a pith ball placed near the conductor of an electrical machine touches and is then strongly repelled. We might also have peeped round the lecture room door one day in January 1836 when Faraday was experimenting with a giant metallic cage, each side measuring 12 ft, in order to determine whether an external electrical charge could be detected inside the cage. The result of this experiment was reported to the Royal Society towards the end of the following year. Usually, however, he conducted his research in the basement where strangers were not admitted. It was here, away from the public, that he performed many of his innovative experiments. However, a close connection existed between his private researches and his public lecturing. He frequently presented to the audience his latest discoveries and his views about both nature and science. Moreover, the beautiful experiments that he performed before his auditors were often large-scale versions of small laboratory experiments that he had devised a few days before the lecture.

Among his fellow scientists Faraday achieved a reputation as a brilliant researcher. Resulting from the long hours spent in the laboratory communing with nature, Faraday produced nearly 400 scientific publications including a series entitled 'Experimental Researches in Electricity'. In the first paper of this series he announced his discovery of electromagnetic induction and he subsequently published the laws of electrochemistry, diamagnetism and the magneto-optical effect. In other papers he announced the discovery of electromagnetic rotations (the principle of the electric motor), the liquefaction of gases and the discovery of benzene. These and other innovations have resulted in Faraday being viewed as *the great discoverer*. As significant as these discoveries were, exclusive concern with them has led to a misunderstanding of Faraday's science. In Chapters 4 and 5 we will discuss several of his leading discoveries, but they will be presented in relation to two other historical themes that shed considerable light on Faraday.

One is Faraday's recurrent concern to comprehend the natural world as a divinely created entity. Thus in his lectures and occasionally in other writings he discussed God's relationship to the physical universe, the nature of matter, electricity and force, and how the different forces of matter constituted an economical system. While such concerns do not feature significantly in the work of present-day scientists, Faraday's science encompassed these issues, and so we should interpret him within a much older tradition, that of *natural philosophy*. Second, we will be concerned with Faraday's laboratory *practice*: the ways his thought and action combined in creating situations in which nature could be incisively analysed and her laws revealed. Although scientists, philosophers and sociologists have proposed many different theories about the nature of science, by observing Faraday in the laboratory we can obtain a far clearer and more accurate notion of scientific practice. Thus a subsidiary aim of the book is to offer some insight into the activity of science.

In the final chapter we attempt to set Faraday and his work within the broader patterns of history. Since Faraday's discoveries form the basis of modern electrical science we can appreciate the long-term influence of his work in relation to such familiar items of electrical equipment as the telephone, the computer and the car generator, to name but three. Moreover, Faraday's ideas profoundly affected many contemporary and subsequent scientists such as Charles Wheatstone, William Thomson (Lord Kelvin) and James Clerk Maxwell, who extended his work into electrical technology and the mathematical theories that form the backbone of modern physics. While Faraday's many discoveries indubitably place him among the most important scientists in history, we should not overlook more subtle aspects of his historical importance. His role as a scientific communicator has already been stressed and to this must be added his role in forging the conceptions of science that we have all inherited.

Note

1. The phrase 'men of science' was widely used in the nineteenth century and is more appropriate than 'scientist' which carries overtones of professionalism. It also reflects the fact that, at that time, very few women were actively engaged in science.

2 From Bookbinder to Scientist

2.1 CAREER

Faraday first encountered science in the books he bound as an apprentice. They introduced him to the fascinating world of natural phenomena and to the idea that these phenomena obey divinely-ordained laws that we can comprehend. He therefore resolved to study science. What is so impressive is the strength of his resolve for during the next few years he thoroughly prepared himself for a career in science at a time when science offered few openings, particularly for somebody without inherited wealth or contacts. He commenced his studies by reading any scientific works he could find and by attending lectures at the City Philosophical Society. He also attended four of Sir Humphry Davy's lectures at the Royal Institution, with tickets donated by a customer at the bindery. Moreover, at the City Philosophical Society (see section 3.2) he met another young man, Benjamin Abbott, and they started a regular correspondence with the aim of *self-improvement*. We see from these letters how Faraday set about improving his mind, his letter-writing skills and his knowledge of chemistry. In the course of this correspondence he commented extensively on the lecturing styles of the lecturers he heard; this critical evaluation of others doubtless paved the way for his own polished performance in the lecture theatre.

In 1812 Faraday completed his apprenticeship and commenced working for a binder whom he found uncongenial. He therefore sought employment in science which he considered a noble activity. As he later recounted, 'My desire [was] to escape from trade, which I thought vicious and selfish, and to enter into the service of Science, which I imagined made its pursuers amiable and liberal.'[1] A letter to Sir Joseph Banks, the President of the Royal Society, brought no reply, so he wrote to Humphry Davy enclosing the bound notes he had taken at his lectures. Davy was obviously impressed by the young man's diligence and enthusiasm. By Faraday's good fortune Davy was involved in a laboratory accident which affected his eyes and, for a time, severely curtailed his activities. He therefore invited Faraday to serve as his assistant while continuing to work in the bindery. Early in 1813 the Royal Institution's laboratory assistant was sacked, following a fight, and Faraday was hired in his place. Davy recommended his new assistant to the Managers (who were responsible for running the Royal Institution) as 'well fitted for the situation. His habits seem good, his disposition active and cheerful, and his manner intelligent'.[2]

After working hard for several months as Davy's assistant, learning numerous skills necessary for laboratory work, Faraday was invited to become Davy's amanuensis during a tour of the continent. For the next 18 months the young man who had not previously travelled outside London journeyed across France, Switzerland, Italy, Germany and Belgium. The tour was a crucial part of his education. As his travel diary and letters record, he observed carefully the landscape, the people and the customs of the regions through which the visitors passed. But the main object of the tour was to enable Davy to meet continental scientists and the travellers spent much time talking to scientists and visiting laboratories in Paris, Rome, Florence and many other cities. While in Paris Davy, with Faraday's assistance, made a significant discovery when he analysed burnt seaweed and identified iodine as a new chemical element. Faraday clearly impressed a number of the scientists they encountered and he subsequently remained in contact with some of them through correspondence. By contrast he came into conflict with Davy's wife, Jane, whom he found haughty and overbearing and she 'endeavour[ed] to thwart me in all my views & to debase me in all my occupations'.[3]

Back in England, Faraday's career prospered. By 1815 he was living at the Royal Institution with the title 'Assistant and Superintendant of the Apparatus of the Laboratory and the Mineral Collection' and with the reasonable wage of 30 shillings a week. In 1821 he was appointed Superintendant of the house and laboratory and he was promoted to Director of the Laboratory in 1825 with a salary of £100 a year. During these early years he worked exceedingly hard assisting the lecturers prepare their demonstrations; he also conducted their research and performed routine chemical analyses as directed. He offered a number of lectures at the City Philosophical Society and began to pursue semi-independent research principally in the area of chemistry. The results of this early research were published in a number of short papers.

Faraday's changing relationship with Davy provides a crucial thread in the study of his career. As Davy's assistant, Faraday was greatly overshadowed by the renowned scientist, yet in order to establish himself in science Faraday had to make an important, independent discovery. He siezed his opportunity in 1821 when the scientific community was greatly excited by Hans Christian Oersted's discovery that when electricity passes through a wire mounted parallel to a compass needle, the needle turns at an angle to the wire. In a review article written in the summmer of 1821 Faraday critically summarised the experimental evidence and theoretical speculations in this burgeoning field of research. He avidly engaged this exciting new research topic with the intention of trying to establish himself by making a significant discovery. Hardly had the ink dried on the first part of his review article when he discovered that a current-carrying wire rotates round the pole of a magnet. He rushed into print announcing these startling 'new motions', usually called electromagnetic rotations. However, no sooner was the paper published than his honour and honesty were challenged. Friends of Davy charged him with failing to acknowledge Davy's contribution adequately and with using the work of another eminent British scientist, William Hyde Wollaston, without acknowledgement. Cut to the quick, Faraday fought to clear his name. However, his relationship with Davy was severely soured.

The matter exploded into the open again in March 1823 when, in front of the Royal Society, Davy was (wrongly) reported as having attributed to Wollaston the discovery of electromagnetic rotations. A few weeks later some of Faraday's friends proposed that he should be honoured by election to the Royal Society and they signed a proposal certificate. However, Davy, then President of the Royal Society, tried to block Faraday's membership. While it is difficult to ascertain Davy's motives, he probably resented his assistant being honoured by membership of the leading scientific society at a time when his own research was waning. Davy was, however, unsuccessful and Faraday's election took place in January 1824. As Faraday wrote many years later, 'I was by no means in the same relation as to scientific communication with Sir Humphry Davy after I became a Fellow of the Royal Society as before that period.'[4] However painful this conflict it marked an important step towards his independence from Davy and his identity as a scientist. Only with Davy's death in May 1829 did Faraday gain the freedom and confidence to pursue an independent scientific career.

Two of Davy's legacies deserve further discussion. The first is that through his experience as Davy's assistant Faraday learnt the many skills of an experimental chemist: blowing glass, constructing apparatus, accurate weighing of samples and titration, to name but a few. These skills remained with him throughout his career and were extended as he engaged new areas of science and developed new, but often impressively simple, pieces of equipment to interrogate nature. At the Royal Institution he had one of the best-equipped laboratories in Britain and he often spent his own money on purchasing new equipment. He was a brilliant experimentalist who effectively combined both head and hand in devising new apparatus and executing novel experiments. Throughout his life he maintained a keen interest in technical processes and machine design; he joined the Society of Arts in 1819 and later served as chairman of its chemical committee which vetted new designs and processes. Moreover, he was in close and frequent contact with instrument makers such as John Newman, who constructed many pieces of equipment that Faraday designed. Thus while Faraday was a highly original thinker he was also a person who possessed immense practical skill.

The second of Davy's legacies was the long-term and time-consuming research that Faraday pursued on behalf of the Royal Society into improving the manufacture of optical glass. Although he presented the results in a lengthy paper in late 1829, a few months after Davy's death, this tedious project only ended in July 1831. It is no coincidence that Faraday's mature and highly creative research began shortly thereafter.

Faraday's *Diary* entry for 29 August 1831 contains the start of his research into electromagnetic induction, which will be discussed further in section 4.3. He was now a self-confident researcher able to follow through an extensive programme of research, creatively combining speculations, inferences and incisive experimentation. The resulting paper, which was read at the Royal Society on 24 November 1831, quickly established his reputation as a first-rate scientist. This paper also marked the beginning of a long series of connected articles with the title 'Experimental Researches in

Electricity' which stretched over a period of 25 years. Some of the research topics will be discussed in Chapters 4 and 5, but it is important to recognise that these closely related research papers constitute the central thread in Faraday's research work. Published in three volumes, they are comparable, in historical importance, with Newton's *Principia* and Darwin's *Origin of Species*.

The late 1820s and early 1830s also marked a change in his lecturing role. Following his appointment as Director of the Laboratory in 1825 he inaugurated the Friday Evening Discourses which became a prominent feature of the Royal Institution's history and a major public vehicle for his own innovations. He soon also began offering his own lecture courses and, in the winter of 1827–8, he delivered his first series of lectures for juveniles. Yet in the late 1820s his extensive duties made Faraday dissatisfied with his post at the Royal Institution, and he explored other posts. However, he considered it 'a matter of duty and gratitude on my part to do what I can for the good of the Royal Institution in the present attempt to establish it firmly'.[5] The words *duty* and *gratitude* show that Faraday felt greatly indebted to the Royal Institution for providing him with an entrée into science and with all the facilities he required. For his part he conceived himself as an employee – a servant – who was duty-bound to act for the Institution's benefit; for example, by ensuring its financial future through his lectures and consultancy work. His salary was augmented by the fees he received, generally about £100 for a course of lectures. In addition he received from the Royal Institution free accommodation, coals and candles. After Davy's death Faraday felt liberated and was far more satisfied with his position. In 1833 he was honoured by being appointed the first Fullerian Professor of Chemistry, a position which was created especially for him and carried a salary of £100 a year.

The contrast with Davy possesses other dimensions. Although Davy was also from a poor background he thoroughly enjoyed the social and economic advantages conferred by success in science. He moved in high society, received a knighthood and a baronetcy, married a rich widow and paid less and less attention to scientific research. What Faraday perceived as Davy's corruption was an object lesson to Faraday who adhered to the biblical injunction, 'Lay not up for yourselves treasures upon earth' (Matthew 6:19). This did not mean accepting poverty but rather living without hoarding money or accumulating luxuries. By the mid-1830s Faraday was earning several hundred pounds a year, the income of a reasonably well established London doctor.

Despite his fame Faraday remained the servant of the Royal Institution. By contrast with Davy he displayed an aversion to power, patronage and politics. Thus he expressed his disapproval of honours, such as knighthoods, because he believed that in Britain they had become debased since they had regularly been awarded for political services and not truly earned. By contrast, he readily accepted the honours bestowed on him by other countries, such as France and Prussia, which he believed operated a meritocratic system. In 1835 he also initially refused a Civil List pension offered ungraciously by the Whig Prime Minister Lord Melbourne but was later persuaded to accept it since, he was assured, he fully deserved the honour. Later he declined the Presidency of the Royal Society, the highest scientific position in the land, which was earlier held by Davy. In 1864 he was greatly disturbed by the offer of the Presidency of the Royal Institution, claiming that 'it is quite inconsistent with all my life & views'.[6] As the Royal Institution's servant, prepared to work unremittingly for its benefit, he could not conceive himself accepting the role of master, the man with political power. Indeed, at that time he even considered whether he and Sarah could continue living in their rooms at the Royal Institution. After 1858 he occupied a grace-and-favour house at Hampton Court.

It is noticeable that Faraday rarely mentioned contemporary social and political events in his correspondence. He believed that politics, in particular, manifested the gross nature of human beings and detracted from the religious ideal of striving for moral perfection. He especially viewed social turmoil with horror and in December 1848, as revolutions were spreading across much of Europe, he complained to a close friend of the 'black passions and motives that seem now a days to urge men every where into action. What incredible scenes every where, what unworthy motives ruled for the moment, under high sounding phrases, and at the last what disgusting revolutions.'[7] To this vision of hell Faraday contrasted not only true religion but also science. While politics was morally regressive and ephemeral, he believed that scientific discovery was a major source of progress and truth for humankind. Moreover, science not only discovers the order in the universe but it is an activity conducive to human peace and order. These were further reasons for the superiority of science and why Faraday was attracted to a scientific career and pursued it singlemindedly.

Faraday's massive research output in the 1820s and especially the 1830s contrasts strikingly with the early 1840s when, for a time, research almost ceased. Moreover, he lectured less frequently at the Royal Institution. The reason for these changes was illness. Although he had earlier complained of various problems, he now suffered from extreme exhaustion, dizziness, headaches and, worst of all, loss of memory. He wrote gloomily to a friend that his doctors 'want to persuade me that I am mentally fatigued and I have no objection to think so. My own notion is, I am permanently worse: we shall see.'[8] He had recovered significantly by 1844 but thereafter often complained of memory problems. It has proved impossible to identify a single cause of this (and other) illnesses. He probably suffered a minor brain lesion in December 1839 but this does not explain all his symptoms. Heavy metal poisoning has been suggested since he worked extensively with lead, mercury and other toxic chemicals. Another possibility is that the 'breakdown' of the early 1840s (and perhaps other occasions) was brought about by unresolved psychological tensions often associated with obsessional personalities. The jury may remain out for a long time to come on this question.

By the mid-1840s Faraday was again pursuing his research and lecturing energetically. He became more prepared to speculate in public and during this period he offered some of his most celebrated Friday Evening Discourses dealing with the the nature of matter and power. In one such discourse, in 1846, he revealed his 'Thoughts on Ray-Vibrations' and speculated that instead of billiard-ball atoms the physical universe is comprised of fields of force. By sprinkling iron filings near a magnet we can plot this field as lines running between the poles. If magnetism can be conceived in terms of lines, so can electricity and gravitation. Moreover, these lines behave like stretched strings which can transmit vibrations across space. Since light was generally thought to be a wave motion, he speculated that these vibrating lines of force constituted light. While Faraday's papers read before the Royal Society and published in its *Philosophical Transactions* were predominantly experimental, his lectures reveal that Faraday was an agile speculator who developed alongside his experimental discoveries a natural philosophy, a view about the constitution of the physical world.

In early Victorian Britain scientists played an increasing role in civil life, bringing their expertise to bear on such topics as safety on the railways, public health and limiting the emissions from factory chimneys. The scientist as expert played an increasingly important role in assisting and strengthening the agencies of central government. Faraday was part of this movement. He was frequently approached by such bodies as the Admiralty, the Home Office and various Royal Commissions to offer his expert, scientific judgement on matters that affected society and about which government legislation had to be framed. We find him contributing to Royal Commissions dealing with gas lighting in art galleries, the site of the National Gallery and the management of harbour lights. In 1844, together with the geologist Charles Lyell, he was sent by the government to contribute expert evidence to a coroner's inquest into the deaths of 95 men and boys in a mining explosion at Haswell, near Durham. Far more demanding was his extensive work for Trinity House on improving the design of lighthouses and their illuminants. Such work was often physically demanding, and we gain a glimpse of its rigours from a letter of 1860, when he was 68 years old, recounting how his train had been caught in an intense snowstorm near Dover and because the road was impassable he had reached the South Foreland lighthouse 'by climbing over hedges, walls, and fields' in order to make 'the necessary inquiries and observations'.[9]

In the early 1850s spiritualism reached Britain from America. In the salons of high society and in working men's clubs spirits moved tables, rapped messages and produced apparitions. The scientific community was divided over the reality of these 'phenomena'. Some considered them worthy of close and scientific scrutiny while others, Faraday included, dismissed the phenomena and considered the mediums charlatans and the public gullible. As a scientist, and therefore an expert, Faraday's views were avidly sought. He therefore carried out a brief, but to his mind decisive, investigation into tableturning and announced his conclusion in *The Times* of 30 June 1853. The phenomenon, he clearly stated, was caused not by spirits but by an involuntary muscular motion produced by the medium pressing on the table. Far from ending the controversy Faraday's statement only increased the volume of mail from the supporters of spiritualism. On 6 May 1854 Faraday attacked spiritualism in a lecture at the Royal Institution before Prince Albert and other notables. The public, Faraday asserted, was gullible because it was not educated in science and could not therefore differentiate between the legitimate claims of science and the false views of the spiritualists. The answer, then, was more scientific education. This was sweet music to Albert, who is said to have embraced Faraday warmly after the lecture. Whatever the importance of science education, Faraday had a more personal reason for attacking spiritualism since he believed that it opposed the Word of God. As we shall now see, religion was crucially important for Faraday.

2.2 RELIGION

Faraday belonged to a small Christian sect called the Glasites or Sandemanians. At no time during its history has the sect's membership exceeded 1000 souls and today only a handful remain. The name Glasites incorporates the sect's founder, a minister named John Glas, who broke away from the Church of Scotland in the 1720s because he considered that the established churches had forsaken the Bible. The small group that assembled round him followed his call for a pure, undefiled Christianity based firmly on directions God had clearly enunciated in the Bible. Moreover, they looked to Christ and tried to follow his moral example. Several Glasite churches were founded in Scotland and Glas's son-in-law, Robert Sandeman (whose family have included the wine importers), assisted their spread to England and America.

Faraday's grandparents, who lived in Clapham, North Yorkshire, first encountered the sect in the early 1760s. Faraday's father, James, subsequently became a member. Members of the sect emphasise the importance of Christian fellowship so, when James fell on hard times, it was arranged for him to move to London and he was subsequently employed in a smithy owned by a Sandemanian. As a child Michael Faraday regularly attended the meeting house near the Barbican. There he met Sarah, the daughter of Edward Barnard who was an elder in the church and a prosperous silversmith. In June 1821 the ironsmith's son married the silversmith's daughter. She was already a member of the church and a month after their marriage Faraday made his confession of faith. This bound him to live according to the sect's strict discipline and in imitation of Jesus Christ. The date of his admission to the sect, 15 July 1821, was undoubtedly the most important day in Faraday's adult life. Apart from a short period in 1844 when he was excluded, he remained in the church for the rest of his life. He also served as a deacon from 1832–40 and as an elder from 1840–4 and 1860–4. Since only those who adhere strictly to the Bible may be elected to the elder's office, his appointment indicates his high moral standing among the congregation.

The Sandemanian church played a crucially important role in Faraday's life. He was required to attend church on the Sabbath and sometimes rushed back to London on the Saturday evening when scientific business took him away from the metropolis. Particularly when he held the elder's office he would preach at the Sabbath service. Surviving copies of Sandemanian sermons show that they consisted almost entirely of passages from the Bible with a minimum of linking material. Thus the word of God was not diluted or distorted by human verbosity. Faraday, like other Sandemanians, possessed an excellent knowledge of the Bible which he read daily. Moreover, he was required to minister to the poor and needy in the congregation. He donated a significant portion of his income to the church and frequently visited and tended the sick. This required a remarkable degree of humility and obedience.

While the Sandemanians shared a common religious outlook, they were also firmly linked by familial ties. Just as Faraday had married into the Barnard family, so there were numerous similar cases of inter-marriage between members of a few leading families. Faraday's younger sister, Margaret, married Sarah's brother, John. Furthermore, just as James had been assisted by fellow Sandemanians in his move to London, so Faraday's life was bound up with other members of the community. For example, in 1848 he recommended a young Sandemanian named Benjamin Vincent for the post of Assistant Secretary at the Royal Institution. Soon Vincent took over the running of the Library, a duty he discharged effectively for nearly 40 years. Again, the group portrait (Plate 1) was painted by a Sandemanian artist, Alexander Blaikley. Another Sandemanian, John Zephaniah Bell, also painted a fine portrait of him (see Plate 2). A further example is that in old age Michael and Sarah were tended by a niece, Jane Barnard, who likewise belonged to the sect.

According to a well-known story Faraday was excluded from the sect in 1844 because he visited Queen Victoria one Sabbath instead of attending the meeting house. However, there is no evidence for this claim. Faraday was certainly excluded, but this arose from problems

Plate 2 J. Z. Bell's portrait

of discipline that were endemic within the church at that time. The exclusion 'brought me low in health and spirits'.[10] Although it lasted only five weeks it placed him in an exposed position since, according to the sect's disciplinary code, one exclusion may be condoned but a second exclusion leads to complete severance from the church and its sanctuary. Moreover, he did not reoccupy the elder's office for another 16 years. He finally laid down the elder's office in 1864, when he was 72 years old, in declining health, and worried by the offer of the Presidency of the Royal Institution.

By the early 1860s his scientific output had fallen dramatically and he was increasingly contemplating 'the future [i.e., eternal] life which lies before us'.[11] He summed up his attitude towards death in a letter to his close friend, the Genevan scientist Auguste de la Rive:

> I am, I hope, very thankful that in the withdrawal of the powers & things of this life, – the good hope is left with me, which makes the contemplation of death a comfort – not a fear. Such peace is alone in the gift of God; and as it is he who gives it, why should we be afraid? His unspeakable gift in his beloved son [Jesus] is the ground of no doubtful hope . . . I am happy & content.[12]

Confirmed in his Christian faith and in eternal life and hope, Faraday died on 25 August 1867. Five days later members of the family and a few friends buried him in the unconsecrated area of Highgate cemetery close to several other Sandemanians' graves.

It is widely believed that science and religion are fundamentally incompatible: the so-called 'conflict thesis'. After all, it has been repeatedly claimed, science is based on reason while religion bows to irrational authority. Such stark contrasts have provided antagonists with much ammunition, but a closer examination shows the relationship between the two important aspects of human experience to be far more complex and interesting. Faraday saw no antagonism between his science and his religion. In a letter to Ada, Countess of Lovelace, he stated that 'the natural works of God can never by any possibility come in contradiction with the higher things that belong to our future existence'.[13] We shall see, below, several ways in which Faraday's religion and his science were in harmony.

While the example of Faraday refutes the conflict thesis, we should not generalise that science and religion are harmoniously related. Faraday turns out to be a very special case for several reasons. For example, his brand of Christianity differed markedly from, say, liberal

Anglicanism and his conception of science also differed radically from that of most of his contemporaries. Again, he confined himself almost entirely to the subjects of electricity, magnetism and chemistry, thus avoiding the increasingly problematic questions raised by contemporary geology and biology (evolutionary theory in particular). The example of Faraday also refutes another well-publicised thesis, that science is compatible only with liberal theology and not with biblical literalism. It should now be clear that, as a Sandemanian, Faraday stood firmly in the latter, not the former, tradition.

Although Faraday told Ada Lovelace that his science and his religion were totally separate and non-interacting, there are many places in his scientific writings where he drew on his religious insights. Most importantly, he conceived the natural world on which he experimented as having been created by God. Thus, as he told an audience in 1846, the properties of matter 'depended on the power with which the Creator has gifted such matter'.[14] The word *power* played an important role in Faraday's mature science since he conceived physical bodies to be a collection of powers. Hence he would have considered this book as possessing the powers of gravity (it is heavy), heat, light and electricity (it might be electrically charged). These powers that consitute the book were created by God at the Creation. Moreover, since the creation and annihilation of power is 'only within the power of Him who has created',[15] the total amount of power in the universe has remained constant since the Creation. Hence force must be conserved in physical processes. As we will see in Chapters 4 and 5 this theologically-based principle had significant implications for his science. Some of Faraday's most important researches were founded on this principle of conservation. For example, he sought the relationship between electricity and magnetism in the belief that a relationship must exist between these forces and that the total quantity of force remains constant. These beliefs prompted his search for electromagnetic induction. Such examples suggest the ways in which Faraday's science and his religion were closely interrelated.

2.3 CHARACTER

Faraday has sometimes been described as a simple man. While he certainly lacked deviousness and cunning, this description implies that he was a rather naive and uncomplicated person. However, closer analysis shows the complexity of his character and some of its opposing strands. Thus he could be both relaxed and driven, intimate towards others and distant from them. Moreover, his public persona differed significantly from his private self (or selves). We will briefly uncover some of these differing aspects of his character, but let us start by seeing him through the eyes of some of his contemporaries.

There can be few people and even fewer scientists who have been so widely admired and loved as Faraday. Hermann von Helmholtz, the eminent German scientist, recalled having received 'the privilege of his obliging help and the pleasure of his amiable society. The perfect simplicity, modesty, and undimmed purity of his character gave to him a fascination which I have never experienced in any other man.'[16] The astronomer Sir John Herschel appended to a portrait of Faraday a Greek tag meaning 'blameless seer',[17] while a contemporary clergyman dwelt in a sermon on Faraday's 'humility and modesty'.[18] These and many other testimonies leave no doubt that he possessed an extraordinarily attractive personality and that he made a deep impression on both close friends and strangers attending his public lectures.

Many aspects of his public persona can be traced to his Sandemanianism which emphasised practical Christian fellowship and the need to treat everybody – Sandemanian and non-Sandemanian alike – with love. Indeed, many of the qualities listed in the above quotations, such as humility, modesty, blameless(ness) and purity, possess strong religious connotations. Faraday tried to deport himself according to the demanding moral precepts of his religion and testimonials show that he was largely successful in this. Moreover, there is a noticeable similarity between the Sandemanian conception of a Christian community dwelling in unity and Faraday's view of the scientific community. In a letter to a friend who was locked in a controversy Faraday referred to the scientific community as a 'band of brothers'[19] and to another correspondent who was frequently the centre of controversy he wrote, 'When science is a republic, then it gains: and though I am no republican in other matters [being a loyal subject of Queen Victoria], I am in that.'[20] Scientists (like Christians) should therefore work together in a spirit of cooperation. This is why he generally stood aloof from the many contemporary scientific disputes and entered into them only when he felt morally bound to clear his name, as against the charge of plagiarism levelled by Davy.

Like Helmholtz, many contemporary scientists (particularly young ones) received his help, encouragement and friendship. Even the children who attended his juvenile lectures were encouraged to gather round the lecture table and repeat the experiments when the lecture was over. Yet, paradoxically, he declined to have any students of his own. Science was a very private avocation and while he would encourage others to study science and pursue research he could share his research activity only with his trusted assistant, Anderson. His strained relationship with Davy made him refuse to place himself in a similar untenable position with a younger person.

Although Faraday could be very gentle he possessed an inner strength that frequently manifested itself. John Tyndall, who joined the Royal Institution in 1853, stated, 'There was no trace of asceticism in his nature. He preferred the meat and wine of life to its locusts and wild honey.'[21] Moreover, while Faraday is generally portrayed as a remarkably warm, empathetic person, he drew a sharp boundary between the acceptable and the unacceptable. Thus contemporaries were sometimes offended when he refused to participate in their political lobbies or to support their advocacy for some entrepreneurial scheme. He was also in some respects a very distant person who inhabited the strange worlds of science and of Sandemanianism. While science provided him with a subject of general interest, his religion imposed an unbridgeable gulf between himself and his non-Sandemanian contemporaries. In Blaikley's picture (Plate 1) the large empty space between Faraday and his audience may be taken to symbolise the considerable distance between the Sandemanian church and contemporary society.

However, there is another side to Faraday's character that emerges only in his correspondence with fellow Sandemanians and a few close friends. He sometimes evinced deep anxieties. For example, he was greatly afraid that he might be a hypocrite: he might not be truthful to himself and to his God but might be living a life of deceit. A further example was his fear that he might be engulfed by the disorder and confusion of the outside world, as exemplified by revolutionary politics. Both these fears, and other similar ones, indicate the tensions that existed behind his calm exterior. These may also be connected with his immense drive. As already emphasised, he worked his way from being the son of a poor blacksmith to the apogee of British science. His main research papers, when collected together, occupy four volumes, while his published *Diary* spans seven volumes. Add to these his numerous lectures and extensive consultancy work and you gain the impression of a very active, busy person. Although he enjoyed such relaxations as music, theatre visits and reading novels, he was acutely aware that the time God had granted him on earth was limited and had to be spent judiciously. Thus in 1832, when his research was progessing apace, he wrote to the Admiralty apologising that he was unable to offer his services freely and added 'but my time is my only estate'.[22] (Faraday did not, of course, inherit a fortune or a landed estate, but was a self-made man.) Likewise he objected to engaging in social frippery because it was time-wasting. As he advised an aspiring young scientist, 'The secret is comprised in three words – Work, Finish, Publish.'[23] Faraday was a highly directed person.

A related aspect of Faraday's work is the order he imposed. God had created the universe according to incorrigible laws and the scientist had a duty to create an ordered environment so as to comprehend those laws. As we noted earlier, Faraday was attracted to science partly because, unlike politics, science was a domain of peace and order. His concern with order manifested itself most remarkably in his *Diary* in which he numbered the main sequence of paragraphs consecutively from 1 to 16 041 over a period of 30 years. Likewise, the 30 series of 'Experimental Researches in Electricity' consisted of 3361 numbered paragraphs written over 25 years and filling three volumes in the collected edition. These sequences of ordered paragraphs produced two remarkable linear structures which recorded on the one hand Faraday's laboratory practice and, on the other, his polished communication with the scientific community. Both sequences presume that scientific knowledge is cumulative, that the results of different investigations will cohere over time and produce a major synthesis of knowledge. To aid this synthesis Faraday made extensive cross-references between paragraphs so as to draw on the wealth of earlier materials. There were also references to the investigations of his contemporaries which he drew into his research but both of these works are principally accounts of how this one man ordered his universe. We will describe his universe in Chapters 4 and 5.

Notes

1. *Correspondence*, letter 419.
2. F. Greenaway, M. Berman, S. Forgan and D. Chilton (eds), *Archives of the Royal Institution, Minutes of the Managers' Meetings, 1799–1903* (15 vols in 7, London, 1971–6) vol.5, p.355.
3. *Correspondence*, letter 46.
4. Jones, vol.1, p.353.
5. *Correspondence*, letter 336.
6. Sarah Faraday to Henry Bence Jones, 31 May 1864: Royal Institution, Faraday papers.
7. *Schoenbein*, p.182.
8. Ibid., p.182.
9. *Reports on the Electric Light to the Royal Commissioners, and Made by Order of the Trinity House, Parliamentary Papers*, 1862, vol.54, p.5.
10. *Schoenbein*, pp.122–3.
11. *Selected Correspondence*, p.1001.
12. Ibid.
13. Jones, vol.2, pp.195–6.
14. Faraday, 'A course of lectures on electricity and magnetism', *London Medical Gazette*, 2 (1846), p.977.
15. 'On the conservation of force', *ERCP*, p.447.
16. *Nature*, 3 (1870), p.51.
17. S. Ross, 'John Herschel on Faraday and on science', *Notes and Records of the Royal Society of London*, 33 (1978), pp.78–92.
18. S. Martin, *Michael Faraday: Philosopher and Christian* (London, 1867), p.24.
19. *Selected Correspondence*, pp.834–5.
20. J. Tyndall, *Faraday as a Discoverer* (2nd edn, London, 1870), p.187.
21. Ibid., p.184.
22. *Selected Correspondence*, p.229.
23. J.H. Gladstone, *Michael Faraday* (3rd edn, London, 1874), p.123.

Plate 3 Exterior of the Royal Institution, from a watercolour by T. Hosmer Shepherd, *c.*1840

3 Faraday in the World

Faraday has been frequently portrayed as a typical lone man of science working unremittingly in his basement laboratory and only when he had made a major discovery announcing it to an appreciative audience in the lecture theatre of the Royal Institution. This image is reinforced by the seven large printed volumes of research notes (referred to as his *Diary*), which contain few references to him working with other scientists. In his work on electricity and magnetism Faraday did indeed work mainly by himself. However, his research notes should not be taken to imply that he spent *all* his time in the laboratory. In fact, only a small proportion of his time was taken up by the discoveries that made him famous and most of his research work was performed in the summer months prior to the commencement of the London social season which directed the calendar of the Royal Institution. The remainder of his time was mostly taken up with fulfilling the duties arising from the various institutions with which he was connected. These duties can be divided into four categories: first, his family and the Sandemanian church; second, commercial consultancy work; third, scientific institutions, and, finally, state agencies and those supported by the state. His involvement in the last two will be discussed in this chapter.

3.1 THE ROYAL INSTITUTION

The scientific institution with which Faraday was most closely associated was the Royal Institution of Great Britain. This had been founded in 1799 by, among others, Benjamin Thompson (Count Rumford) and Joseph Banks (President of the Royal Society) with the aim of 'diffusing the Knowledge, and facilitating the general Introduction, of Useful Mechanical Inventions and Improvements; and for teaching, by courses of Philosophical Lectures and Experiments, the application of Science to the common Purposes of Life.'[1]

It rapidly became a fashionable centre for the dissemination of scientific knowledge through lectures, most notably those of Humphry Davy who had lectured there from 1801. Also, mainly through Davy's work on the electrochemical isolation of sodium and potassium as chemical elements, the Royal Institution had established an international reputation for research.

In March and April 1812 Faraday attended the final four lectures that Davy delivered at the Royal Institution. Faraday transcribed these neatly and, soon after his apprenticeship was over, presented them to Davy, requesting a position at the Royal Institution and hence employment in science. He was interviewed by Davy who advised him to 'Attend to the bookbinding',[2] but added that if an opportunity should arise he would think of Faraday. In late October Davy sustained a serious eye injury when a combination of azote (nitrogen) and chlorine exploded in his face. Being prevented from using his eyes Davy employed Faraday temporarily as his amanuensis.

On Friday, 19 February 1813, William Harris, the Superintendent of the Royal Institution, heard a 'great noise' in the lecture theatre. On investigation he found William Payne, the Laboratory Assistant, and John Newman, the Instrument Maker, engaged in a bitter row. Newman charged Payne with neglecting his duty by failing to attend the Professor of Chemistry, William Thomas Brande. Newman also complained that Payne had hit him, a charge upheld by the Managers at their meeting on 22 February. Payne was promptly sacked. Within a week Davy had asked Faraday if he would be willing to take over as Chemical Assistant. According to a later recollection by a friend, Faraday was preparing for bed when the message arrived from Davy asking to see him the following morning. After a further interview with Davy, he was appointed Laboratory Assistant on 1 March 1813. In October Faraday agreed to join Davy on a tour of the Continent for which a special passport had to be obtained from Napoleon since the two countries were still at war. Following their return to England in April 1815, Davy was appointed a Manager and a Vice-President of the Royal Institution and, at his first meeting with the new board of Managers, it was decided that Faraday should be re-engaged in his former position and that he should also be granted accommodation in the building.

Faraday remained the Chemical Assistant for the next ten years, during which time he assisted various lecturers at the Royal Institution with their demonstrations and also delivered lectures on chemistry to medical students and others who needed to learn the subject. Early in 1826 he was relieved of the task of assisting other lecturers with their demonstrations. Between 1815 and 1826 Faraday also helped Davy with his research, particularly on developing the miner's lamp and on the utility of attaching copper sheathing to the bottom of ships. There were, however, breaks from this routine – notably when Davy was abroad or touring in Britain – when he treated Faraday as his London agent, asking him to transact business on his behalf or to check a chemical analysis. He even invited Faraday to join him in Naples to assist with some chemical work, but Faraday declined.

In 1821 Faraday was appointed Superintendent of the House and, in 1825, Director of the Laboratory. In 1828 his value to the Royal Institution was acknowledged by an invitation to attend the Managers' meetings. His duties were also changing. From the beginning of 1826 Faraday embarked on expanding the Royal Institution's programme to disseminate scientific knowledge. This programme was all the more necessary since the Royal Institution had, once again, fallen on hard times. It is clear from many contemporary letters that Faraday considered it his duty to help alleviate this position. He founded the Friday Evening Discourses in 1826 and the Juvenile Christmas Lectures in the following year. The increased lecture programme attracted new members and ensured the continuing support of existing ones. Their subscriptions helped rescue the Royal Institution from its financial difficulties.

These lecture series helped raise the visibility of both Faraday and the Royal Institution. From their inception the Friday Evening Discourses were reported in detail in the Saturday newspapers, such as the *Literary Gazette* and the *Athenaeum*, and less frequently in *The Times*. The Juvenile Lectures were also widely reported in the press. These reports served to bring science before the educated general public and to spread Faraday's reputation beyond the scientific community to a much larger appreciative audience. The public visibility of both science and the Royal Institution was particularly important to Faraday. Thus, in the late 1830s, he also initiated the construction at the Royal Institution of an imposing corinthian columned façade (see Plate 3). This façade distinguished 21 Albemarle Street from the adjacent shops and residential buildings and proclaimed that it alone was devoted to the pursuit of learning.

Until the early 1840s Faraday organised the Friday Evening Discourses singlehanded, although he was ultimately responsible to the Lecture Committee. Besides delivering many of these Discourses, he invited the speakers and occasionally had to step in at short notice when a speaker let him down. He also arranged displays of objects in the Library. These were sometimes connected with the subject of the evening's Discourse, but more often than not they reflected the diverse cultural interests of the audience at the Royal Institution. These articles ranged from exhibits that Stamford Raffles (the founder of Singapore) brought back from his expeditions to a Leigh (of Leigh–Enfield) rifle; and from plants and portraits to poems. Faraday also delivered 17 series of Christmas lectures between 1827 and 1860. His last series, devoted to the chemical history of the candle, proved so popular that it has been reprinted regularly and translated into such languages as Polish and Japanese.

Following a breakdown in his health in 1840, the routine administration of lectures passed to others. However, he occasionally persuaded his scientific friends and colleagues to lecture at the Royal Institution. One indication of Faraday's reputation is that he persuaded the Master of Trinity College, Cambridge, William Whewell, to deliver one of the very few lectures Whewell gave outside Cambridge. This lecture formed part of a series on education to which Faraday contributed his famous lecture on mental education. This lecture, and indeed the whole series, arose from Faraday's horror at the public's enthusiasm for spiritualism and tableturning in the mid-1850s. Strongly opposed to these fashions on both religious and scientific grounds, Faraday took every opportunity to publicise his belief that if the public were better educated in scientific matters they would not be taken in by such baseless fancies.

Faraday continued to deliver Friday Evening Discourses until the early 1860s. His final lectures attracted audiences of about 800. Blaikley's portrait (Plate 1) shows a packed lecture theatre. Although Prince Albert was present, it was Faraday the audience had come to see and hear. It is an indication of Faraday's success in establishing this lecture series and the Juvenile Lectures that both continue to flourish today, more than 160 years after their foundation. Indeed the latter, now known popularly as the Christmas Lectures, reach an audience of millions through the medium of television. Faraday's success stemmed from his perception that lecturing was a crucially important part of his scientific work. He did not simply deliver the results of his laboratory researches but carefully thought out how to present scientific knowledge to an audience which included many who possessed no scientific expertise. For Faraday knowledge was not to be kept in his basement laboratory or transmitted only to other men of science. It was, he believed, possible to communicate even the most abstract and technical knowledge to a general audience. But he also realised that his auditors had to be led from simple experiments, which anyone could appreciate, progressively through to the most refined ideas. Even these were judiciously illustrated by experiments which had to be visible to the entire audience. Contemporary reports indicate that his audience were stimulated and learnt a great deal from attending his lectures.

The Royal Institution was not only the place where Faraday was employed to lecture and administer: it was also his, and Sarah's, home. Moreover, it housed his laboratory, which was the best in Britain and one of the best in Europe. The Royal Institution provided Faraday with research, lecturing and domestic facilities that existed nowhere else in Britain. The corinthian columns of 21 Albemarle Street hid an ideal environment for his scientific researches.

3.2 OTHER CLUBS AND SOCIETIES

At the end of his 1872 biography, John Hall Gladstone listed what he thought were all the learned societies to which Faraday belonged. Over 70 learned societies were listed ranging from the most learned and prestigious British, European and American ones to such humble local societies as the Hull Philosophical Society and the Sheffield Scientific Society. In many instances honorary membership was conferred on him. The subjects to which these societies were devoted covered an immense range, from architecture and medicine to geology and pharmacy. This range illustrates that Faraday's reputation went far beyond the confines of his chosen disciplines of chemistry and natural philosophy.

Gladstone's list, which is not complete, is based on Faraday's diploma book, now in the archives of the Royal Society. Many of these societies elected Faraday to membership partly in order to confer honour on themselves. However, a few societies played a central role in Faraday's life. These were the City Philosophical

Society, the Royal Society of London and, to a lesser extent, the Athenaeum, the Society of Arts and the British Association for the Advancement of Science. By examining Faraday's involvement and changing attitudes towards these institutions we can discern changes in his outlook towards secular society.

The City Philosophical Society

The City Philosophical Society, founded in 1808, emerged from chemical lectures given by the silversmith John Tatum. These were open by subscription and were intended to provide artisans and apprentices, like Faraday, with access to scientific knowledge. Faraday first attended these lectures in 1810 by paying a shilling per lecture which his brother Robert gave him. In the early nineteenth century there were many such lecturers not only in London but in many provincial towns and cities. Their audiences were attracted to science partly because it promised better career prospects as Britain continued its industrialisation. Moreover, it was widely accepted that science led to both mental and moral improvement. These were all themes that attracted Faraday and ones which he later stressed in his lectures.

The members of the City Philosophical Society took themselves very seriously and emulated the well-established learned societies. Thus they held formal elections to membership, organised by the Secretary, and used the letters MCPS to signify membership. Every Wednesday evening they met at Tatum's house, 53 Dorset Street. Initially meetings alternated between Tatum lecturing on chemistry and a member giving a lecture on the subject of his choice; subsequently, the format alternated between a lecture by a member and a discussion of that lecture. In 1813 Faraday was actively seeking election to the Society, but this did not take place because of his visit to the Continent with Davy. Shortly after their return in 1815 he was elected and there delivered his first lectures on scientific topics. Indeed, he gave 20 such lectures between 1816 and 1819. Of these 13 were on chemical subjects, five dealt with the properties of matter and in the remaining two he discussed philosophical topics, particularly the acquisition and retention of knowledge.

Through the Society Faraday received much of his early scientific education and he was initiated into the art of lecturing. Some of these lectures included demonstrations, although it is not clear whether

Faraday performed any. This was also the place where he formed some life-long friendships, notably with Benjamin Abbott, Edward Magrath and Richard Phillips. In these respects the City Philosophical Society exerted a profound formative influence on Faraday.

The Royal Society of London

The Royal Society is the oldest learned society in Britain and the oldest such society currently in existence anywhere in the world. It was founded officially in 1660 although its origins go further back. Nowadays Fellowship is restricted almost entirely to distinguished scientists. However, in the early nineteenth century the Society functioned more like a London club and many who joined had only a passing interest in natural knowledge and its applications. A great number of Fellows had Admiralty connections or were members of the landed gentry. In 1820 the 42-year presidency of Sir Joseph Banks ended with his death. He was temporarily replaced by William Hyde Wollaston who was rapidly succeeded by Davy (Plate 4) in November 1820. Davy sought to balance the interests of the various conflicting groups within the Society, but he ultimately failed, resigning in 1827 two years before his death.

Following Faraday's discovery of electromagnetic rotations and the liquefaction of gases, he was an obvious candidate for Fellowship of the Royal Society. With Davy as President he should have had little difficulty in being elected. His friend Richard Phillips, who had only recently been elected, originated Faraday's nomination (at Faraday's suggestion) in May 1823 but without, it seems, consulting Davy as was customary. On 30 May Faraday noted that Davy was angry about his nomination, but did not say precisely why. Davy then asked Faraday to take down the nomination certificate, but he replied that only his proposers could do this. There are a number of possible reasons for Davy's behaviour. As noted in Section 2.1 Davy appears to have resented the scientific successes of his assistant. Another reason was that the outside world, whose opinion Davy courted and valued, might infer that he was continuing the President's tradition of patronage. Banks had been perceived as misusing his position as President in this way and had included a large number of Fellows who could not be called men of science. Davy was anxious not to be seen to use the same sort of patronage as Banks had done. Moreover, he wanted to reduce the annual intake of

Plate 4 Humphry Davy

new Fellows by excluding non-scientific candidates in order to make the Society more scientific. This damaged the interests of some groups within the Royal Society. Why, then, did Davy oppose the election of a rising scientist like Faraday? If he had not opposed his election then many would have assumed he was for it, and so perceive him to be continuing the misuse of patronage. Nevertheless Faraday was elected on Thursday, 8 January 1824. At what must have been an emotionally-charged meeting the following Thursday, Davy admitted Faraday to the Fellowship.

This row also revived the controversy over the accusations made in October 1821 concerning Faraday's originality in the discovery of electromagnetic rotations. Davy was reported as having repeated the accusation that in his paper on rotations Faraday had used the electromagnetic work of Wollaston without due acknowledgement. This report, which Davy denied, forced Faraday to make public his authorship of the 'Historical Sketch of Electromagnetism' and to write his 'Historical Statement' describing the path of experimentation by which he arrived at his discovery. Thus ended Faraday's close personal relationship with Davy.

Their professional relationship did not end, however. As President of the Royal Society and as Faraday's superior at the Royal Institution Davy possessed sufficient power to exercise considerable influence over Faraday's work and career. He changed from being Faraday's benevolent patron to utilising his abilities on various projects without, it seems, worrying if such tasks would be detrimental to Faraday's career. For example, as President of the Royal Society, Davy arranged for Faraday to undertake research into improving the quality of glass for telescopes and other optical instruments. This is a long and involved story with many political ramifications both in the Royal Society and in the government. In 1825 Faraday was appointed to a committee together with the Cambridge-educated John Herschel and the optical instrument maker George Dolland. Faraday was to supervise the manufacture of the glass samples, Dolland was to grind them into lenses and Herschel was to establish their optical properties. By 1827 little progress had been made and the Royal Society and Royal Institution decided, at Faraday's suggestion, that a furnace should be constructed at the Royal Institution to enable him to make the glass samples himself. Almost every day throughout 1827 and 1828 Faraday was involved with this work. Only after Davy's death in 1829 was

Plate 5 Deputation consisting of Lord Wrottesley, J. P. Gassiot and Sir W. R. Grove inviting Faraday to become President of the Royal Society in 1858. Group portrait by E. Armitage

Faraday in a position to abandon this onerous task and return to his own research. So far as Faraday was concerned the only benefits this work brought were that his assistant, Sergeant Charles Anderson, who had been hired for this project, became a permanent assistant in Faraday's laboratory and second, that one of the glass samples made at that time was used by Faraday in his discovery of the magneto-optical effect in 1845.

After Davy relinquished the Presidency Faraday's relationship with the Royal Society improved markedly. He served on the Society's Council in the late 1820s and early 1830s and submitted his major series of research papers, entitled 'Experimental Researches in Electricity', for publication in the Society's *Philosophical Transactions* (the leading British scientific journal of the day). However, Faraday soon began to withdraw from an active role in the affairs of the Society. One reason was the defeat of John Herschel in the 1830 contest for the Presidency. Faraday supported Herschel, whom he considered represented the working scientist, against the Duke of Sussex, who stood principally for aristocratic privilege. By the mid-1830s he had distanced himself considerably from the Society's administration. Despite this he was offered the Presidency in 1858 (see Plate 5) but declined, no doubt remembering Davy's term of office, with the comment, 'if I accepted the honour which the Royal Society desires you [Tyndall] to confer upon me, I would not answer for the integrity of my intellect for a single year.'[3]

The Athenaeum

Outside the Royal Society Davy secured Faraday's services in founding the Athenaeum, a London club which many Fellows of the Royal Society joined. Early in 1824 there had been an exchange of correspondence between Davy and the First Secretary of the Admiralty, John Wilson Croker, in which a new gentlemen's club was proposed. Then, as now, the Athenaeum was to cater for the elite among such groups as scientists, artists and clergy who were not adequately catered for by existing London clubs. Faraday was invited to be its first Secretary. Between March and June 1824 he sent out invitations, compiled lists and so on. Indeed, organising the initial membership took much of his time. Here again we see Davy using

Faraday's talents for his own purposes. Faraday undertook these chores voluntarily and when payment was offered he passed the job on to his friend Edward Magrath. Faraday remained a member of the Athenaeum until 1851 when he resigned, ostensibly owing to his declining income.

The Society of Arts

Faraday was elected a member of the Society of Arts for the Encouragement of Manufactures (now dignified by the prefix Royal) in 1819. In subsequent years he was joined by a number of the other members of the City Philosophical Society as that society disbanded. Although he remained a member of the Society of Arts throughout his life he was most active in the 1820s and 1830s when he served on several occasions as chairman of the Chemistry Committee. In this position he was called to consider the merits, or otherwise, of new chemical processes and their application to industry. Those who were successful were rewarded by a prize or a premium from the Society. For its part the Society of Arts held Faraday in high esteem and, in 1866, he was awarded the Albert Gold Medal for his discoveries in electricity, magnetism and chemistry which had been widely and successfully applied in industry.

Most interestingly Faraday campaigned in 1824 for Joseph Chater, from a Sandemanian family, to be elected as the Society's Collector (the person responsible for collecting membership dues). He wrote to influential members of the Society, such as chairmen of committees, canvassing their support. This he obtained, and Chater was elected by an overall majority. His support of Chater contrasts with his general attitude towards supporting applicants for scientific posts. Faraday claimed that he *never* wrote testimonials for anyone, no matter how well he knew them. He even refused to add his support for the young James Clerk Maxwell. It appears that his dislike of the competitive element prevented him from supporting men of science for various positions. On the other hand, he was willing to vouch for the characters of fellow Sandemanians and the upright members of Sandemanian families. Thus Faraday was instrumental in enabling several Sandemanians to occupy various positions at the Royal Institution.

British Association for the Advancement of Science

The British Association for the Advancement of Science was founded in 1831 as an annual peripatetic meeting of men of science in provincial towns throughout Britain. Then, as now, one of its leading aims was to educate the public in scientific matters. Although Faraday did not participate in the founding of the Association, and was unable to attend the first meeting in York, his interest in popularising science and in meeting other men of science led him to play a role in the Association, although not a very prominent one. He attended about half the annual meetings and was honoured by being elected Vice-President on three occasions and President of the Chemical Section twice. Rarely, however, did he attend the whole of a meeting but often left on the Saturday so that he could be present at the Sandemanian meeting house on the Sabbath.

3.3 FARADAY'S ROLE IN CIVIC AND MILITARY SCIENCE

With the increasing industrialisation of Britain throughout the nineteenth century, government departments and other semi-official bodies found themselves in need of expert advice from established scientists on various technical problems. Faraday was called in to advise many such bodies either on an occasional basis or as a long-term paid adviser. In this section we shall examine some of his work for the Admiralty, the Home Office and Trinity House, but we must not forget that he also did consultancy work for the Board of Trade, the Office of Woods and Forests, the Board of Ordnance and other such bodies. Even his teaching at the Royal Military Academy should be viewed in this context.

It was the Admiralty, perhaps the largest industrial organisation in the world at that time, that was most severely affected by rapid technological change. The advent of the marine steam engine and the introduction of copper sheathing to protect ships' bottoms are two prime examples of how technology altered naval practice. Since the seventeenth century the Royal Society had provided the Admiralty with advice and a close association had grown up between the two. By 1820 the scale and complexity of the technical problems had increased considerably. However, during the 1820s the relationship between the Admiralty and the Royal Society deteriorated considerably. Davy was perceived, perhaps unfairly, by the Admiralty and the general public, as having offered unsound advice on the copper sheathing of ships. Moreover, in 1828 the Board of Longitude – the only formal link between the Royal Society and the Admiralty – was abolished. In the following year the Admiralty set up its own Resident Scientific Committee whose members were Thomas Young, Edward Sabine and Faraday. Young died soon after and Sabine appears to have contributed little to the Committee. Faraday, however, was active in giving the Admiralty advice when requested. In the 1830s this ranged from analysing copper – a favourite concern of the Admiralty – to showing that a consignment of oats had been contaminated with chalk. In the 1850s Faraday advised on the use of telegraphy for naval purposes and, once again, on the problem of copper sheathing during the Crimean War.

In 1829 the Royal Military Academy secured his services as Professor of Chemistry. The Academy was located at Woolwich – it moved to Sandhurst and amalgamated with the Royal Military College in 1939 – and its main function was to train cadets for the Royal Artillery and Royal Engineers. Because of the increasingly technical nature of warfare the cadets received extensive training in science. In the summer of 1829 the post of Professor of Chemistry became vacant and Lt-Col. C. W. Pasley, who had attended Faraday's chemical lectures at the Royal Institution, wrote to the Commandant of the Academy, Col. Percy Drummond, recommending Faraday as the ideal candidate. As he commented, 'Faraday ... is not only one of the best chemists of the day, but certainly the best Lecturer, qualities not always combined.'[4] After some negotiation with Drummond, Faraday was appointed Professor of Chemistry in December 1829. His contract specified that he had to deliver 25 lectures a year for which he would receive £200. These favourable terms indicate how anxious the Academy was to obtain his services. By contrast the Professor of French, who was appointed at the same time, was paid only £150 per year for much longer hours and on condition that he reside at Woolwich. The deal gave Faraday considerable economic freedom from the Royal Institution, and enabled him to dispense with much of his consultancy work. On the other hand, for half of each year from 1830 to 1851 he spent the best part of two days a week at Woolwich. But this was a light load compared with the effort he had previously put into the industrial research projects at the Royal Institution. On taking up the post at Woolwich he relinquished much of this industrial research.

That Faraday spent two days a week at Woolwich for one hour of lecturing indicates the care he took in preparing his lectures. As he noted during negotiations with Drummond, chemistry lectures, unlike mechanics lectures, had to be prepared individually each time they were given because the chemicals involved could not be preserved from one course to the next. Furthermore, as he told Drummond soon after his appointment, 'I think experimental lectures owe all their value to the experiments & visual illustrations which are given in conjunction with the theoretical details & it will be my object to make these demonstrations as distinct and impressive as possible.'[5] We do not know the content of Faraday's lectures to the cadets but they were probably similar to his lectures at the Royal Institution, since he used the same textbooks. There may, however, have been greater emphasis on the military uses of chemistry, as in explosives.

These lectures at the Academy show that, no matter to whom Faraday was lecturing, he took inordinate pains to prepare his lectures. To regard lecturing as mere hack work would have been insulting to his audience and thus contrary to his religious disposition. Unlike some other religious groups the Sandemanians were not inhibited from supporting the armed forces. Thus as a loyal citizen Faraday felt it his duty to provide the Admiralty and the Army, through the Royal Military Academy, with scientific expertise. These lectures at Woolwich were the only ones (besides those at the London Institution) which Faraday delivered outside the Royal Institution. Their effect on British military attitudes towards science is not clear. However, the topic was considered important enough for the Academy to appoint another experienced chemist, F. A. Abel, as Faraday's successor.

On 28 September 1844 there was a mine explosion at Haswell Colliery in County Durham. Ninety-five men and boys were killed. Since there was disagreement at the inquest over who should inspect the mine, James Graham, the Home Secretary, asked Faraday, the geologist Charles Lyell and a surveyor of mines named Samuel Stutchbury to carry out the inspection. After their visit Stutchbury claimed that the accident was due to an accumulation of 'fire damp', a highly inflammable gas, which had been ignited accidentally. The coroner's jury found this evidence sufficient to return a verdict of accidental death. Despite playing only a minor role at the inquest, the presence of Faraday and Lyell conferred high scientific status on the verdict and Faraday subsequently delivered a discourse at the Royal Institution on how science could improve the safety of mines. Thus Faraday was able to serve the civic authority, as Sandemanianism required, but in this case by using his scientific reputation as well as his expertise.

Since 1514 the Elder Brethren of Trinity House have been charged with maintaining safe navigation round the shores of England and Wales. Their task has been to supervise pilots and to build and maintain lighthouses and lightships. Faraday was appointed their Scientific Advisor in 1836. During the Second World War most of the records of Trinity House were unfortunately destroyed, so we know little of Faraday's advice to them. We know, however, that in the 1850s and 1860s Trinity House embarked on a series of tests to examine the practicality and efficiency of replacing oil and gas lights with electric ones. Faraday was heavily involved with these tests and even as an old man of 70 he responded enthusiastically to the call to examine the lights at South Foreland and Tynemouth. This occasionally involved him going out to sea in a small boat in order to estimate the range of the lights. His work on this project was his last major contribution to applied science. In 1865 he arranged for John Tyndall, his successor at the Royal Institution, to take over his duties at Trinity House.

3.4 FARADAY IN THE WORLD

We began this chapter by suggesting that Faraday is often portrayed as a lone researcher who stood apart from the rest of the world. However, we have seen that this is an inaccurate view. Faraday spent much of his life engaged in activities which, though not related to his researches, were closely connected with the concerns of contemporary society. As a Sandemanian Faraday felt a strong sense of duty towards his employer, the Royal Institution, and towards the civic authorities and their agencies. Thus we find him working hard on the optical glass investigation over several years. However, he complained that this work wasted his time for not only did it prevent him from pursuing his own research but he also came to believe that optical glass manufacture could not be improved. To continue with this project was thus a waste of the time that God had granted him.

After 1829, freed from Davy's influence and no longer completely dependent economically on the Royal Institution, Faraday was able to choose which research topics to pursue. He believed that science should be applied for the benefit of mankind. However, he was not a narrow utilitarian since he considered that the scientist's principal aim was to discover the laws which God had framed in constructing the physical universe. Even in its application science possessed a theological message, and in an 1858 lecture on the telegraph he claimed that we should apply science to practical problems in order to convey the gifts of God to humankind. But he also desired to be seen applying contemporary science to practical problems: hence his work for the Admiralty, Trinity House and the other agencies described above. For Faraday science was not to be kept in the laboratory, but should be moved out into the wider world. He was therefore active in those organisations which furthered this aim.

Notes

1. F. Greenaway, M. Berman, S. Forgan and D. Chilton (eds), *Archives of the Royal Institution, Managers' Minutes*, vol.1, p.1.
2. *Correspondence*, letter 30.
3. J. H. Gladstone, *Michael Faraday* (3rd edn, London, 1874), p.55.
4. *Correspondence*, letter 399.
5. Ibid., letter 427.

Plate 6 Faraday's laboratory at the Royal Institution. From a watercolour by Harriet Moore

4 From Chemistry to Electricity

In previous chapters we described the contexts in which Faraday lived and worked. We now turn to examine his experimental work and the world-views that informed it.

4.1 CHEMISTRY

With Faraday's great contributions to physics, it is often overlooked that he was initially interested in chemistry and was trained as a chemist. In Faraday's youth electricity was regarded as part of chemistry, following Volta's invention of the battery which produced electricity by chemical means. The great German organic chemist Justus von Liebig commented:

> I have heard mathematical physicists deplore that Faraday's records of his labours were difficult to read and understand, that they often resembled rather abstracts from a diary. But the fault was theirs, not Faraday's. To physicists, who have approached physics by the road of chemistry, Faraday's memoirs sound like an admirably beautiful music.[1]

This implies that Faraday pursued his physical studies using the chemical methods learnt while an apprentice chemist. Our analysis in this and the next chapter bears out Liebig's judgement.

From the earliest evidence it is clear that Faraday was concerned with issues at the cutting edge of chemical research. His early letters to Benjamin Abbott show him trying to convince a somewhat sceptical Abbott of the correctness of Davy's view of what was called oxy-muriatic acid gas. This was generally thought to be a combination of oxygen and some other element(s), but Davy argued and offered evidence to prove that there was no oxygen in this substance and that it was a chemical element: that is, it was incapable of being decomposed chemically or mechanically. He named it chlorine. Hence oxy-muriatic acid contained no oxygen. This overturned the definition of acidity propounded by French chemists, notably Lavoisier, which stated that acids must contain oxygen. Davy discussed this work in the lectures Faraday attended in the spring of 1812. Faraday supported Davy's interpretation in his early letters to Abbott. That chemistry offered the potential to make such exciting and fundamental discoveries must have been a major stimulus for Faraday to pursue chemistry and to approach Davy for patronage.

Compared with other sciences chemistry ideally suited Faraday's inclination and abilities. Unlike French physics of the time, it was non-mathematical; it contained theories which went beyond the mere collection of data, as occurred in natural history; and it involved active experimentation on natural phenomena. Above all, it was at the leading edge of scientific discovery in England. This view of chemistry was reinforced for Faraday on his Continental tour with Davy, when he was able to witness and take part in a discovery almost on a par with Davy's work on the elemental nature of chlorine: Davy's proof of the elemental nature of iodine. On this Faraday commented in a letter to Abbott: 'They [the French chemists] reason theoretically without demonstrating experimentally, and errors are the results.'[2] For Davy and Faraday, as for most British chemists, experiment was essential to the practice of chemistry. Theory must not become dogma but should be submitted to the rigorous test of experiment and, if it failed, theory must be modified or abandoned. It was this experimental attitude that Faraday developed early in his training as a chemist and carried over to his study of electricity and magnetism.

Of course he had more to learn not only about science, but also about the power of experiment. After returning to England in 1815 Faraday spent several years continuing his scientific apprenticeship. This mainly involved preparing materials for and assisting lecturers at the Royal Institution. In particular, he assisted the Professor of Chemistry, William Thomas Brande. Brande is a somewhat neglected figure; his distinguished reputation, particularly as a textbook writer and lecturer, has been eclipsed by that of his assistant. Because he and Faraday worked so closely together there is very little evidence to indicate how they interacted. However, early in 1826 Brande recommended to the Managers that in view of Faraday's avocation, he should be relieved of assisting him. Unlike Davy, Brande had no wish to inhibit Faraday's career.

During this period Faraday undertook chemical analysis of water samples obtained from different regions of Britain and also helped Davy with his researches. The most noteworthy of these was Davy's invention of the miners' safety lamp and his subsequent work on the theory of flames in 1815–16. Faraday's considerable assistance was duly acknowledged in Davy's papers.

Faraday's early original chemical work included the analysis of caustic lime from Tuscany, which formed the subject of his first paper, his discovery of two carbon chlorides, and his confirmation (with Richard Phillips) of the existence of a third. His most important chemical discovery was the liquefaction of gases. On 5 March 1823 he was analysing chlorine hydrate for Davy, who suggested that it should be heated in a sealed tube. The hydrate decomposed and the pressure increased to four or five atmospheres. This was sufficient to liquefy the chlorine. He then found that by increasing the pressure of chlorine in a tube, he could cause it to liquefy directly. The following day Davy liquefied hydrogen chloride and Faraday repeated the process with many other gases. It is interesting to note that, apparently unknown to themselves, they were repeating work done some years earlier by Thomas Northmore, an obscure Devonshire poet. It is indicative of Faraday's reputation that he is regarded as the discoverer of the liquefaction of gases.

Much of Faraday's chemical research stemmed from his consultancy work. For example, in the early 1820s Faraday did much work with James Stodart, a member of the Royal Institution, trying to improve the quality of steel. They made different alloys. When Stodart died in 1823 the project came to a halt. Their work does not appear to have exerted any influence on the subsequent development of metallurgy, not because of the inherent quality of the steel they produced, which was high, but because it was not commercially viable to manufacture at that time.

Faraday's involvement in various legal cases was important. As Britain industrialised, entrepreneurs as well as government departments needed to know how to cope with the implications of technological change. An increasing number of lawsuits – often involving patents or insurance claims – required expert scientific witnesses. Faraday was called as an expert witness in several such trials but he seems mostly to have been on the losing side! The precedents for expert witnesses were in the process of being constructed so there were no clear procedures to be followed in court. Men of science often appeared for both plaintiff and defence, causing judges and juries much confusion and bringing science some bad publicity.

In 1818 Faraday was asked to determine the originality of Daniel Wheeler's 1816 patent for a new process of preparing malt for beer which involved heating it to 400°F. The defendant, named Maling, had used Wheeler's method after a patent had been granted and so Wheeler sued for infringement. However, it emerged that he had not filed the specification for the process within six months of the date of the patent application. A temporary injunction was thus filed against Wheeler preventing him from enforcing his patent. Wheeler requested that the injunction be nullified. By this time Maling had engaged two chemists as expert witnesses to testify that the process had been in use before the date of the patent. Wheeler employed Faraday to contest this evidence, which he did after visiting Maling's factory and examining his apparatus. Faraday then sought to undermine the evidence produced by Maling's witnesses but to no avail, since the Lord Chancellor continued the injunction.

The most celebrated case, which set precedents for the use of expert witnesses, was that of the sugar baking firm of Severn and King against various insurance companies. The Imperial Insurance Company engaged Faraday as a witness. They refused to settle Severn and King's claim – ultimately for £70000 – following a fire on 10 November 1819 in which their premises were destroyed. The central question was whether a new process involving oil-gas, introduced by Severn and King without the knowledge of the insurers, increased or decreased the risk of fire. Faraday conducted an extensive series of experiments to show that the risk of fire was increased. However, the case went against the insurers. Faraday then continued with these experiments for the Phoenix Insurance Company, but the same judgement resulted.

While Faraday was working on oil-gas he became intrigued by some of the substances formed during its decomposition. One of these he found to be a compound of hydrogen and carbon which he named bicarburet of hydrogen. This was renamed benzene in 1834 by Eilhard Mitscherlich, the German chemist who first carefully studied its properties. Apart from his major work on electrochemistry and some interesting ideas on catalytic action, this was Faraday's last major research in pure chemistry. However, he wrote his only book, as opposed to collections of papers or lectures: *Chemical Manipulation* was published in 1827 and passed rapidly through two further editions. It was a textbook of chemical procedures written for students with no previous knowledge. As such it was atypical among contemporary textbooks. Instead of describing the properties of chemical elements and their compounds, it showed how to perform experiments with them. It can also be read as a compendium of the chemical procedures, techniques and methods that Faraday had learnt during his training at the Royal Institution.

From chemistry Faraday also learnt of the existence of electricity. As noted above, electricity was at that time more closely connected with chemistry than with physics. Faraday, more than any of his contemporaries, took electricity from chemistry and exploited it in the domain of physics. From his chemical apprenticeship he learnt not only the importance of experiment but also the manipulative methods and tacit knowledge that would enable him to explore electromagnetism so successfully. Fifteen years' experience as a chemist proved an excellent training.

4.2 ELECTRICITY, MATTER AND MOTION

Of the discoveries mentioned in earlier chapters the most significant for Faraday was electromagnetic rotation. In September 1821, while exploring Oersted's discovery that an electric current affects a magnetised needle, Faraday made a current-carrying wire move continuously around a magnet. This was the first time continuous motion had been produced from chemical energy. In Figure 4.1 we see the final two pages of his notes for a day's work on the effect of currents on magnetised needles that led to the first rotation apparatus, sketched in the corner. These notes trace his exploration of the Oersted effect, beginning with the search for clues he suspected others had missed, and leading to experiments that did not work, including one to test Wollaston's idea that the wire should rotate on its own axis. We saw earlier that this discovery embarrassed first Faraday, as to its originality, and then Davy, when Faraday's priority was again disputed in 1823 in connection with his election to the Royal Society. In order to establish himself as a scientist, Faraday was eager to make a discovery of his own. The rotations

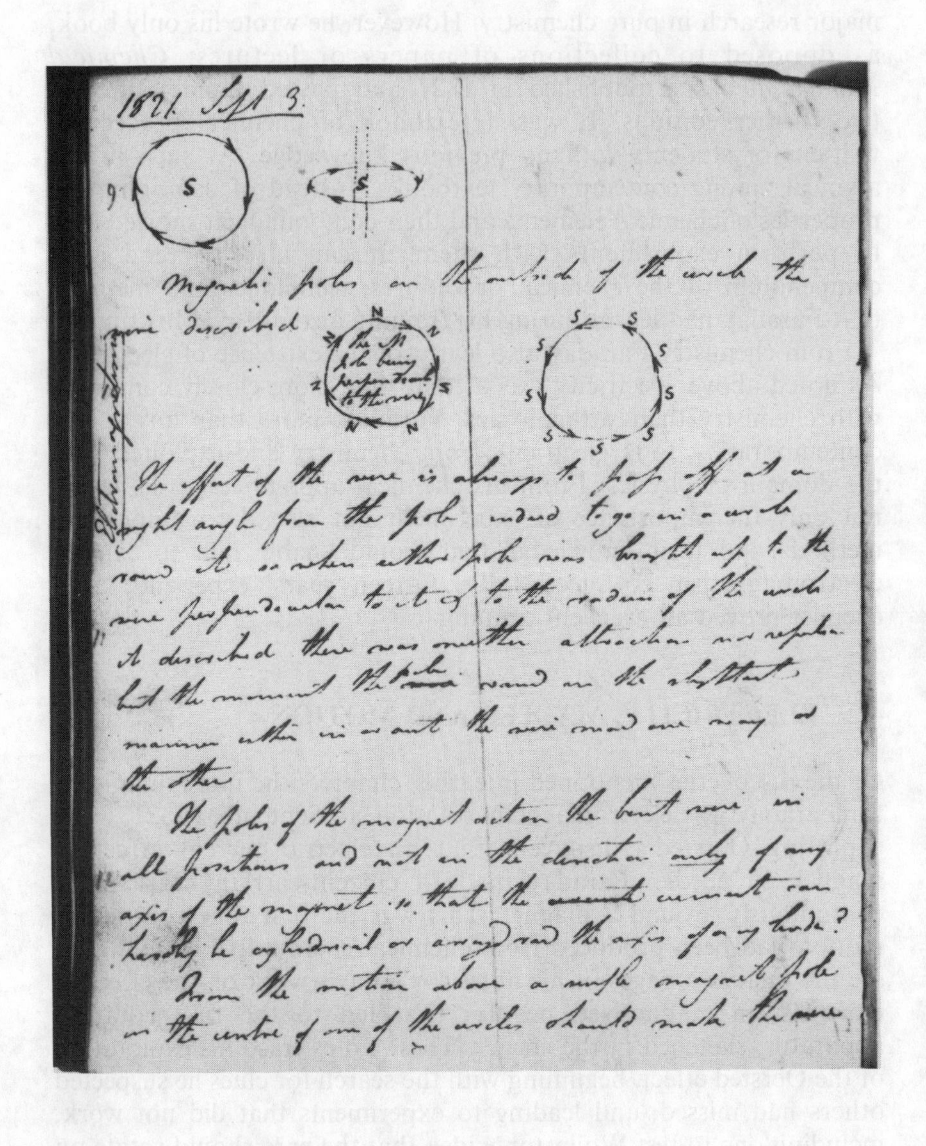

1821. Sept 3.

Magnetic poles on the outside of the circle the wire described

the N pole being perpendicular to the wire

Electromagnetic motions

The effect of the wire is always to pass off at a right angle from the pole indeed to go in a circle round it so when either pole was brought up to the wire perpendicular to it & to the radius of the circle it described there was neither attraction nor repulsion but the moment the pole varied in the slightest manner either in or out the wire moved one way or the other

The poles of the magnet act on the bent wire in all positions and not in the direction only of any axis of the magnet so that the currents can hardly be cylindrical or arranged round the axis of a cylinder?

From the motion above a single magnetic pole in the centre of one of the circles should make the wire &

Figure 4.1 Electromagnetic rotations. The final two pages of Faraday's notes on experiments performed on 3 September 1821 that led to the first electromagnetic motor.

Source: Royal Institution, Faraday Papers.

1821. Sept. 3

continually turns round. Arranged a magnet needle in a glass tube with mercury about it and by a cork water &c supported a connecting wire so that the upper end should go into the silver cup & its mercury & the lower move in the channel of mercury round the pole of the needle. The battery arranged with the wire as before. In this way got the revolution of the wire round the pole of the magnet. The direction was as follows looking from above down

S

N

Very satisfactory but make more sensible apparatus.

Electromagnetism

Tuesday Sept. 4

Apparatus for revolution of wire & magnet. A deep basin with bit of wax at bottom & then filled with mercury. A magnet stuck upright in wax so that pole just above the surface of mercury. Then piece of wire floated by cork at lower end dipping into mercury & above into silver cup as before & [illegible] by wire as carefully attached from being the [illegible]

S

N

showed his ability to do important research in an exciting new field. Besides being Faraday's first independent contribution to electromagnetism, they were the first brushstroke in a new portrait of nature that would develop over the next 30 years. In the remainder of this chapter and the following one we describe how, with one discovery after another, Faraday constructed a radical alternative to traditional views of force and matter.

Faraday's approach to electrical phenomena on the expanding frontiers of chemistry had its roots in his chemical apprenticeship with Davy and Brande, rather than in the Newtonian, mechanical approach of many mathematically educated natural philosophers. Newtonian mechanics made matter and force primary, force being what produces a change in velocity or direction of motion of a body. Faraday did not interpret electromagnetic rotations as forces of this kind. To him they were an instance of the conversion of a more general source of power from one form into another: he believed, as did Davy and others, that the chemical powers of the materials in the battery are electrical in nature, and assume a current or 'dynamical' form in the wire that interacts with the magnet, producing motion. Such phenomena show the expenditure of force in one of its many forms so they should be approached descriptively, qualitatively and, above all, taken at face value. In order to analyse these phenomena mathematically, Ampère and Biot interpreted them as actions produced by Newtonian forces. Faraday, however, rejected reinterpretations which, he believed, adapted phenomena to fit a preconceived theory. His approach was shaped as much by his distrust of mathematics as it was by his chemical apprenticeship.

His distrust of mathematics was not, as is often supposed, a symptom of his lack of scientific training: such training did not exist even in the two universities England possessed at that time. Faraday was – as Maxwell saw – a rigorous thinker, capable of reasoning from one state of a physical situation to the next, reconfiguring each state in his imagination, in a systematic and incisive way. This ability is essential to the qualitative reasoning on which mathematical analysis depends and also for effective thought-experiments. Faraday was so keen to take issue with those who bent phenomena to fit mathematical methods that he threw down the gauntlet by naming the first of his projected series of research papers 'experimental'. This first paper announced the discovery of electromagnetic induction: that is, the induction of an electric current by magnetism.

4.3 ELECTROMAGNETIC INDUCTION

Oersted had shown that electricity produces magnetism and most scientists expected a converse or reciprocal effect, hoping to make electricity from magnetism. Many, including Faraday, tried and some produced the effect without recognising it. But Faraday persisted long after most others had abandoned the search and in the process he learnt what to look for. On 29 August 1831, Faraday connected one pair of contacts of a wire coil *A* in Figure 4.2, wound around a soft iron ring, to a galvanometer or current detector. The other pair of wires from a second coil (*B*) he connected to a battery. The needle of the detector moved and then returned to its rest position, indicating that a current had passed briefly. Faraday made this

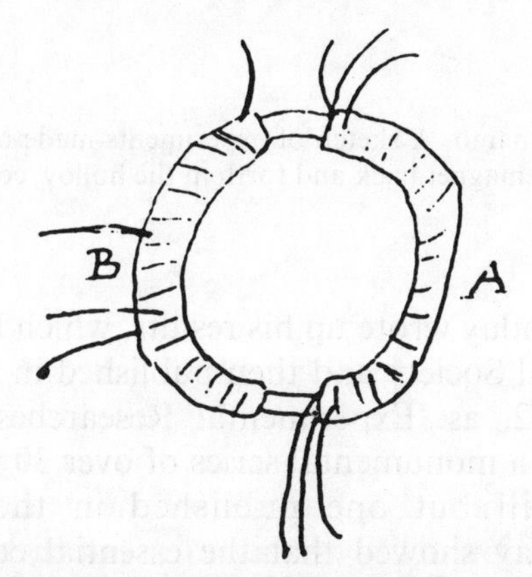

Figure 4.2 Electromagnetic induction. Faraday's sketch of the soft iron ring wound with wire coils which he used to produce electricity from magnetism on 29 August 1831.
Source: *Diary*, vol.1, p.367.

elusive effect visible because he made one crucial variation in the sequence of procedures: he connected the coil to the current detector *before* he made the battery connection. This is one of the favourite discovery stories in the history of science. The first arrangement was a primitive transformer: the current in coil *B* induces magnetism in the iron ring, and this magnetism, in turn, induces a current in coil *A*. Faraday soon elaborated this basic effect, showing that a moving magnet induces a current in a coil as in a simple dynamo (Figure 4.3). The rotations of 1821 showed that electromagnetism could produce motion. By demonstrating that magnetism in motion can produce electricity, he had laid the foundations of electrical technology.

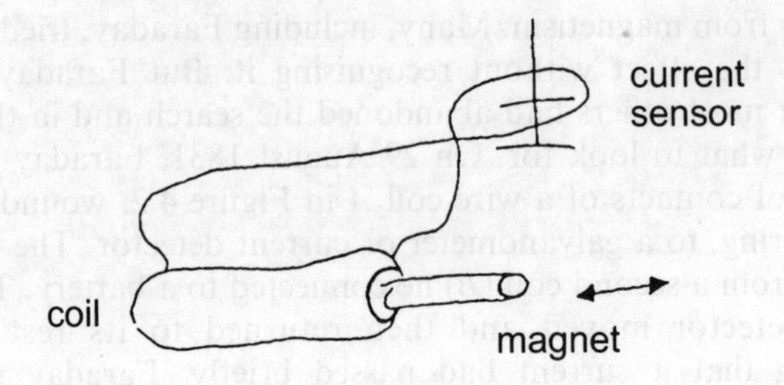

Figure 4.3 The first dynamo. A sketch of experiments made on 17 October 1831. Moving the magnet back and forth in the hollow coil induced a current in the coil.

In November Faraday wrote up his results, which he first read to a meeting of the Royal Society and then published in its *Philosophical Transactions* in 1832, as 'Experimental Researches in Electricity'. This was the first of a monumental series of over 30 papers spanning nearly 25 years, all but one published in the *Philosophical Transactions*. Faraday showed that the essential condition for the magnetic induction of a current is that the conducting circuit must cut through the system of lines or 'curves' depicting the magnetic force emanating from a magnet or another current. Figure 4.4 shows his first published sketch of these lines. Faraday had not used magnetic curves in earlier papers, but the imagery was familiar enough. Other natural philosophers had used this convention to describe magnetic phenomena by analogy to lines of latitude and longitude and to the isothermal lines that represent the distribution of temperature of the earth.

First introduced as aids to the memory, then used as a descriptive concept, Faraday's lines became a powerful theoretical tool which William Thomson and James Clerk Maxwell subsequently found particularly useful. Faraday was soon using the lines in several ways: as a descriptive and interpretative tool to analyse new phenomena, to

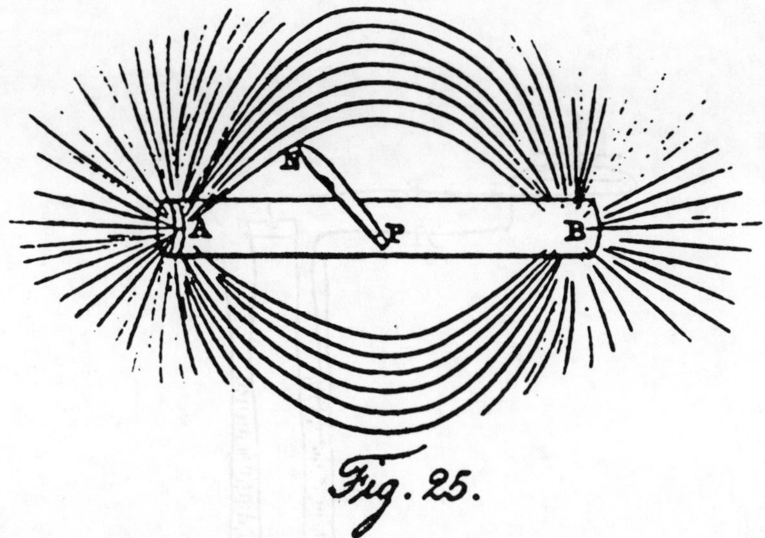

Figure 4.4 Magnetic curves. Faraday's first published drawing of the curves or lines of magnetic action surrounding a bar magnet.

Source: *Researches*, vol.1, plate 1.

explain known effects and to make new possibilities for further experiment. His discovery of unipolar induction illustrates how a thought experiment with lines led to a new experimental situation. By imagining a system of lines emanating from a magnet as a reference frame extending from one pole to the other, and supposing that as the system rotates its curves are cut by a wire positioned halfway up the magnet – as shown in Figure 4.5 – he inferred that a current would be induced in the wire. This was the first time electricity had been generated by unipolar induction. When described in terms of 'curves', electromagnetic induction could also explain other phenomena such as the rotation of a copper disc between the poles of a magnet when current was supplied to it. Since its discovery by François Arago in 1825, others had tried unsuccessfully to explain the wheel's rotation mathematically. As Faraday wrote to a friend, his discovery and application of electromagnetic induction showed that 'experiment need not quail before mathematics, but is quite competent to rival it'.[3]

These and other discoveries encouraged a flurry of speculation; Faraday believed he was on the brink of a major breakthrough. Two speculations are particularly important because they anticipate ideas he later developed into a coherent theory of the magnetic field. On 12 March 1832 he deposited a sealed note with the Secretary of the

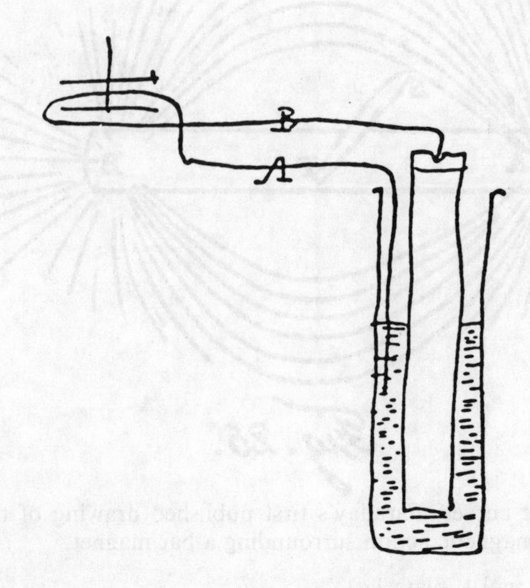

Figure 4.5 Unipolar induction. The magnet was positioned vertically with its lower end half immersed in mercury. Rotating the magnet about its own axis generated a current through a circuit consisting of the wire, mercury, current detector and the magnet itself.

Source: *Diary*, vol.1, p.403.

Royal Society. In it he suggested that the magnetism which induces currents 'proceeds gradually from the magnetic bodies, and requires time for its transmission', and that the induction of electrostatic charge 'is also performed in a similar progressive way'. Faraday thought that the mode of transmission in each case would be like 'the vibrations upon the surface of disturbed water, or those of air in the phenomena of sound' and suggested that 'the vibratory theory will apply to these phenomena, as it does to sound and most probably to light'.[4] Was he already thinking in terms of electromagnetic waves?

A second speculation in his laboratory *Diary* for 26 March 1832 shows Faraday trying to combine three different aspects of electromagnetic induction in a single, dynamic process. There he pointed out that the 'mutual relation of electricity, magnetism and motion may be represented by three lines at right angles to each other',[5] where action by any two will produce the third along the third line. His sketch (Figure 4.6) shows that he already possessed a general model for the interaction of electricity, magnetism and mechanical force.

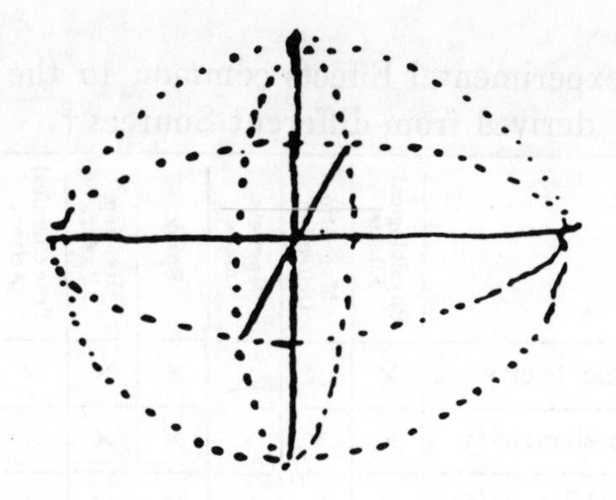

Figure 4.6 Faraday's sketch of 26 March 1832 showing the general relationship between three physical quantities: Electricity, magnetism and motion.

Source: *Diary*, vol.1, p.425.

4.4 THE IDENTITY OF ELECTRICITIES

Spurred on by his success, Faraday was eager to demonstrate experimentally the identity of electricities. At this time the different electricities were still named according to their sources: static electricity produced by friction was called 'frictional' or (because of its familiarity) 'common' electricity; current produced by a chemical pile or voltaic cell was 'voltaic electricity' and, since Volta's discovery had originated in the twitchings of the legs of Galvani's frogs, there was also 'galvanism' or 'animal' electricity. The uneven heating of two metals produced 'thermo-electricity' and Faraday's magnetically-induced currents were called, of course, 'magneto-electricity'. He believed that all these electricities are forms of one and the same power, but this idea was not self-evident and had to be justified. He prepared his case, as he had prepared for electromagnetism in 1821, by studying the literature on the subject, looking for evidence in other people's work and repeating many experiments. His method of proof was chemical. He embarked on a systematic analysis of the properties of phenomena traditionally attributed to different electricities. These properties were the effects each kind could exhibit: heating, magnetism, chemical decomposition, physiological effects, and so on. Faraday summarised his results in the table shown in Figure 4.7.

Table of the experimental Effects common to the Electricities derived from different Sources †.

	Physiological Effects.	Magnetic Deflection.	Magnets made.	Spark.	Heating Power.	True chemical Action.	Attraction and Repulsion.	Discharge by Hot Air.
1. Voltaic electricity	×	×	×	×	×	×	×	×
2. Common electricity...	×	×	×	×	×	×	×	×
3. Magneto-Electricity..	×	×	×	×	×	×	×	
4. Thermo-Electricity...	×	×	+	+	+	+		
5. Animal Electricity...	×	×	×	+	+	×		

Figure 4.7 Identity of electricities. Faraday's table showing the properties displayed by electricity derived from different sources. X indicates properties found by January 1833. He added the +'s later, to indicate properties he had established by December 1838.

Source: *Researches*, vol.1, p.102.

When publishing his results Faraday presented the identity of electricities as a self-contained scientific issue. However, as we have seen, it was much more than that. Such findings were entrances into the study of the unity and connectedness of every aspect of God's creation. Faraday believed his duty as a scientist was to disclose the unity within God's plan. But there were always new findings to accommodate. As he experimented, Faraday produced new effects which demanded explanation. For example, to determine whether the discharge of static electricity has magnetic effects, Faraday tried to 'slow down' the discharge by placing wet string in the circuit, so that it might have time to affect his galvanometer. The water made a big difference. When he held the electrodes, the water seemed to absorb the force, reducing the shock delivered to his arms and chest. (In this and other instances, he used himself as a detector, to prove the 'animal' or physiological effects of static electricity.) Could matter absorb and momentarily store electricity? He also showed that static electricity can effect chemical change. As yet he could not explain facts such as these. They gradually convinced him that matter cannot be divided into two distinct classes, conductors and 'electrics' (insulators), and that electricity is not a material substance.

4.5 ELECTRICITY AND MATTER

The established theories of electricity hypothesised either one or two quasi-material fluids which were thought to reside on or in bodies, rather as water is absorbed by paper but forms a thin film or beads on a non-absorbent surface, like steel. The fluid became visible when moving between bodies, as sparks. Whenever he mentioned these fluid hypotheses in a lecture, Faraday would sternly remind his audience that they were mere suppositions and that the nature of electricity was not known. But the many new properties of electricity discovered between 1790 and 1831 had made its nature an increasingly pressing problem. Chemists had known since 1800 that a voltaic current can decompose water. Current electricity – measured roughly by the number and size of the plates of a battery – quickly became an important investigative tool for chemists. Many, including Davy, concluded that the chemical affinities which hold the particles of substances together are actually electrical. Just as static electricity can be discharged as a spark, so voltaic current was, they thought, a converted and mobile form of chemical affinity. This analogy was incomplete: in the first place, current is produced by chemical change but the matter of the reagents combines to form new substances; it does not turn into electricity. Second, a current stronger than the one produced by given reagents can reverse their reaction, but it is not converted into more reagents.

Faraday wanted to know how electricity and matter are related. To examine their relationship he pursued a detailed and systematic study of electrochemical phenomena, experimenting with various properties of the current and of reagents. During the first half of the nineteenth century chemistry was very important to Faraday's colleagues at the Royal Institution because it possessed so many commercial, maritime, military and agricultural uses. It is therefore not surprising that he approached the study of matter through electricity and both in turn by chemical methods. Since electricity is intimately involved in chemical change, the study of electrochemical phenomena was an appropriate way to advance on two fronts at once.

From 1833 Faraday concentrated on electrochemical decomposition. An early discovery surprised and perplexed him. When the solution in a cell is frozen, both electrical conduction and decomposition cease. Faraday saw this as an instance of a general rule, that matter (the solution) loses or gains certain powers (such as conductivity) when it changes state. He first tried to explain the dependence of conductivity upon physical state in terms of inter-particle forces, but soon abandoned the attempt. His many experiments on decomposition yielded his two laws of electrochemistry, two different theories of electrochemistry, a new vocabulary for electrochemistry and, of great importance for the development of his own ideas, they raised further difficulties about the relationship between electricity and matter.

In his search for a clearer conception of this relationship Faraday pursued the quantitative relationship between electricity and chemical change. He showed in 1833 that the amount of chemical action caused by a current passing through a solution is directly related to the quantity of electricity passed, and that the masses of the products evolved, deposited or dissolved is proportional to their chemical equivalents. In order to show this he solved many problems, ranging from the design of special collecting and measuring instruments to the study of secondary effects that interfered with measuring the products of the primary actions which these instruments were meant to collect. One of the most important of these enabling studies concerned the effect of the metal platinum, which he used as one of the poles of his cells. Faraday found that even very clean platinum induced chemical changes on its surface when no current was passing, and that other metals could act in the same way if sufficiently clean. His study of catalytic action (so named by Berzelius in 1836) is one of many occasions when experiments produced new phenomena but offered few clues about how to explain them. Faraday proposed an elaborate theory of inter-particle forces to explain catalysis, but soon abandoned it. By the spring of 1834 he was complaining that he still had no clear ideas about how electricity and matter interact. He replaced his first theory of electrolysis by a more spartan theory, based upon four general assumptions about chemical force. These explained his laws of electrolysis (something the earlier theory had not done). However, his enduring achievement was a new electrochemical vocabulary. Most of his terms – such as ion, anion, cation (for the reagents), electrode, cathode, anode and electrolyte – are still used today.

How were these new terms invented? Let us take the 'poles' as an example. Chemical products are precipitated or evolved as gases at the metal poles of a cell. These were called poles because many believed that they attract or repel particles of reagents: the poles at which hydrogen and oxygen appear when water is decomposed were named, respectively, the hydrogen pole and the oxygen pole. But Faraday doubted that poles act on particles; neither were they simply the places where chemical products appear. He thought of them as gateways, the places where current could enter and leave the solution, so as to complete the circuit. The current enabled chemical change by allowing inter-particle tensions to be released: it was an 'axis' of power along which tensions and discharges moved. (We shall see this idea reappear towards the end of his career, as a general definition of electric and magnetic lines.) Since chemical products are material, this axis should also be a pathway for matter. But matter could not simply be identified with force: to say this would raise far larger questions than experimental philosophy could deal with. The current both enables and is caused by chemical change throughout the whole solution and it has a definite direction.

Faraday devised this new descriptive language with the help of William Whewell of Trinity College, Cambridge, a keen scientist and accomplished classicist. Faraday described the entities and situations the new words should describe and Whewell suggested words derived from Greek roots. To tie in with terrestrial magnetism and lines of latitude, Faraday gave Whewell an example based on an east–west current. Instead of 'pole', Whewell suggested the words 'eisode' and 'exode' (meaning 'a way in' and 'a way out') and 'anode' and 'cathode' (meaning an 'eastern' and a 'western way'). Faraday eventually chose anode and cathode. This collaboration with Whewell was so successful that many of their terms are still used in electrostatics and magnetism as well as electrochemistry.

Why did Faraday need new words? There were no new phenomena to name (as there had been with magneto-electricity or the new rotations) and the familiar language of atoms, forces, positive and negative fluids and poles was better understood by his readers than new terms would be. He also knew that 'names are one thing and science another'.[6] Here Faraday confronted the double-edged nature of language: words and images give us the power to describe possibilities beyond our experience. However, familiar, established words can also constrain the way we think. Faraday wanted to say something new about familiar phenomena. This meant freeing them from the theories implied by the traditional language of electrochemistry. To do this he had to redescribe the phenomena so as to free them from habitual ways of thinking about them. As he explained to Whewell in April 1834, he felt that the prevailing ideas about electricity were 'very clumsy' and had 'little doubt that the present view of electric currents and the notions by which we try to conceive of them will soon pass away'.[7] In a letter to C. Lemen he expressed his suspicion that 'the usual notions attached to Positive and negative and to the term current . . . are altogether wrong but I have not *a clear view* of what ought to be put in their places',[8] so he wanted 'names by which I can refer to [things] without involving any theory of the nature of electricity'.[9] Faraday needed to clear a space for a new theory, even though he could not yet say what this was.

A '*clear view*' of how electricity and matter interact continued to elude Faraday. He could not explain puzzling phenomena such as why, when a liquid freezes, it not only ceases to conduct electricity but can even become charged. This fact had impressed him so greatly that he could not refrain from suggesting that 'all bodies conduct electricity in the same manner from metals to [shel]lac and gases, but in very different degrees'.[10] Here a chemist boldly contradicted one of the most cherished and established assumptions of electrostatics: that all material substances fall into two physically distinct classes, conductors and insulators. Instead of drawing this firm line, Faraday supposed that electricity acts identically on all substances. But to make his case, the non-chemical (molecular) forces activated in changes of state would eventually have to be reintroduced. Here he was confronted by the difficulty of reconciling the fact that electricity appears to have two states – static and dynamic – with the assumption that matter is, or must be, particulate and not a continuum. He could not achieve this reconciliation. Although his published papers hint at the need for a fundamental rethink of the problem, he abandoned particulate or atomistic theories altogether by the spring of 1834.

Notes

1. A.W. Hofmann, 'The Faraday Lecture', *Journal of the Chemical Society*, 13 (1875), p.1100.
2. *Correspondence*, letter 49.
3. *Correspondence*, letter 522.
4. *Selected Correspondence*, p.217.
5. *Diary*, vol.1, p.425.
6. *Researches*, vol.1, p.198.
7. *Selected Correspondence*, p.264–5.
8. Ibid., p.267.
9. Ibid., p.264.
10. *Researches*, vol.1, p.125.

Plate 7 Attracting the world: Michael Faraday holding a bar magnet

5 From Electricity to Natural Philosophy

5.1 SELF-INDUCTION

Faraday's next major discovery strengthened his doubts about traditional ways of conceptualising active agents, like matter and electricity. In 1834 he learnt of a new electro-magnetic effect. A strong aftershock was felt both on breaking a circuit containing a coil with a soft iron core, and on breaking a circuit consisting of a very long wire. In the former case the aftershock was due to an induced current, but at first Faraday thought it was produced by a change in the magnetisation of the soft iron core when the current was broken. As for the second case, it appeared that the current acquired a greater intensity by passing through an increased length of the circuit. An obvious explanation was to suppose that the current is a fluid which, like water in a pipe, acquires more momentum as it passes through a greater length of wire. This explained, for example, the fact that a spark jumps when a circuit is broken. It also implied a mechanical explanation of electrical phenomena which Faraday did not want to accept. If, as he now believed, an electric current releases tensions between chemical affinities, then a tension should exist prior to the passage of a current. He found that he could draw sparks on closing a circuit (*before* a current is flowing in it) as well as on breaking a circuit. This showed that such tensions exist. It cast doubt on the idea that the sparks drawn on breaking a circuit are caused by the electric fluid's momentum forcing it across the gap. The momentum-explanation takes the materiality of electricity for granted, and it flew in the face of the transfer of magnetic effect from one circuit to another. Electric sparks were at least visible, but how could a substance move *imperceptibly* from one wire to another?

Faraday used lines of magnetic action to redescribe both cases. He then argued that both effects have a common cause in the magnetic action of the current which, when the current is broken, induces a second current in the conductor. This description enabled further experiments. These demonstrated the inductive nature of the effect because, he thought, fluid theories could not account for them. For example, he showed that when a current is broken (*a* in Figure 5.1), a current can be made to appear in a neighbouring (or secondary) circuit. Placing a second circuit (*b*) next to the primary (*a*), he found that sparks which normally appeared at the gap in *a* appeared instead at the gap in *b*. This mutual induction showed that 'the *extra current* could be removed'[1] from *a* to *b*. Removing *b* restored the sparks to *a*, showing that the secondary current is induced in whichever conductor happens to be present. Faraday used lines of magnetic force to

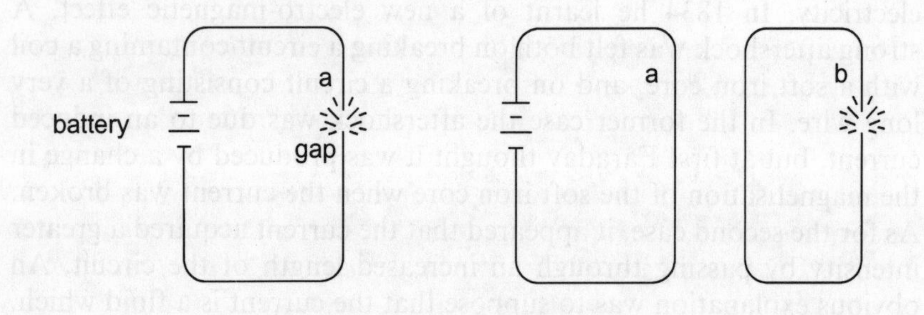

Figure 5.1 Self- and mutual-induction. A spark due to self-induction occurred at the gap in circuit *a*. When a second circuit, *b*, was placed alongside *a*, the spark occurred in the gap in circuit *b*. This showed that the inductive effect of circuit *a* could be transferred to circuit *b* and that self- and mutual-induction are the same

argue that all of the inductive action that would normally affect the primary circuit is transferred to the secondary, illustrating this by using the number of lines of induction to represent the quantity of inductive action received by the secondary circuit. He rejected the idea that the primary current 'circulates with something like *momentum or inertia* in the wire'[2] and by December 1834 he had concluded that all the effects are due to 'the induce current described . . . in the First Series of these Researches'.[3] His use of magnetic lines to express the transfer and conservation of inductive action anticipates what Thomson and Maxwell later called the energy of an electromagnetic field, but the concept of energy still lay two decades in the future.

5.2 ELECTROSTATIC INDUCTION

Although Faraday was moving ever further from the accepted ways of theorising about electricity and matter, he had so far articulated only the beginnings of an alternative view. A fresh approach was needed to deal with the new phenomena discovered during the previous half-century, including the voltaic pile, electrochemical and electromagnetic phenomena. By 1835 he realised that he would have to work out the physical implications of his approach and compare it with traditional theories of electrostatics and magnetism. In 1837 and 1838 he published a vigorous experimental challenge to the whole of classical electrostatics. How did Faraday arrive at this position?

His challenge originated in doubts about the distinction between conductors and insulators. As we have seen, he was puzzled by new facts such as the non-conductivity of electrolytes when they freeze. The new questions posed by electrochemistry convinced him that a general theory of matter–force interactions would have to wait for a better understanding of the relationship of electricity to the particles of matter. Faraday expected a lot from the new theory. Beginning in August 1834 he made lists of key problems in his *Diary*. These show that he was 'thinking much lately of the relation of common [static] and voltaic electricity: of induction by the former and decomposition by the latter'.[4] In November 1835 he noted the need 'to make out the true character of ordinary electrical phenomena'. He recorded that he was 'convinced that there must be the closest connexion'[5] between induction and different forms of conduction, and that electrical philosophers had missed the truth. He put this possibility as a question: are insulators really necessary to the existence of charge? This question would have made no sense to adherents of fluid theories of electricity.

Faraday then considered the following problem: if the tension *within* the electrolyte is sustained and continued by lines of induction *outside* a chemical cell, through the space between the electrodes (Figure 5.2a), then the charged state – whether of a chemical battery, the bulb of an electrostatic generator or the polished metal globes of a condenser (or Leyden jar, as in Figure 5.2b) – cannot be the accumulation of a quasi-material entity. An even more radical possibility occurred to him. As early as May 1834 he noted in his *Diary* that 'it may be possible to have a current of Electricity without

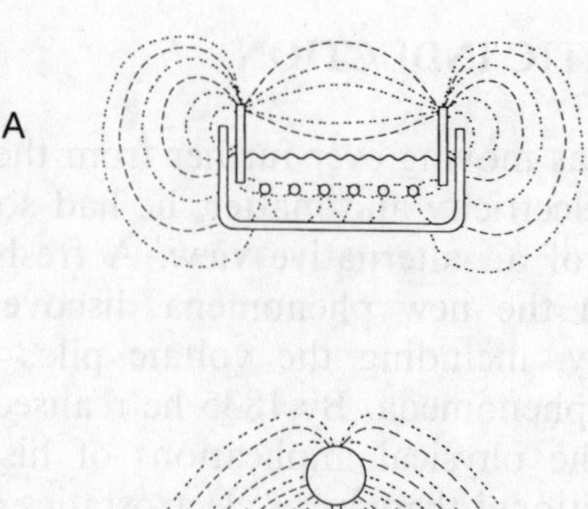

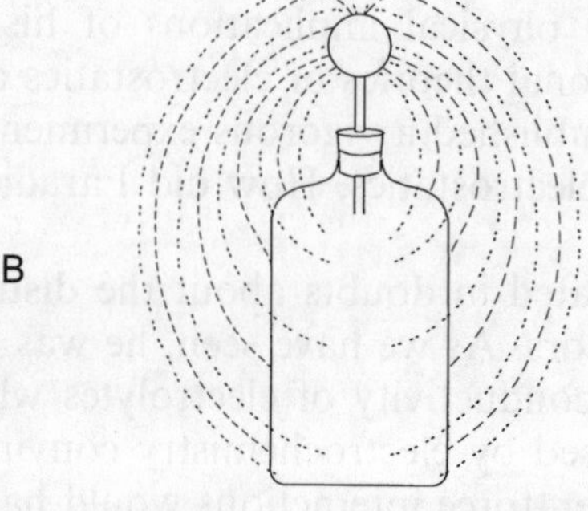

Figure 5.2 Electricity and matter. Faraday's lines of electrostatic induction enabled him to visualise the analogy between a chemical battery (*A*) and a charged condenser or Leyden jar (*B*).

a circuit, that is, to have an absorptive effect at each end of a series of apparatus. It would be a current between two vessels not forming a circle'[6] (i.e., a closed circuit). The theory he expounded in 1837 is based on an analogy between the polarised state of the matter of an 'electric' (or insulator) in the region of charged conducting surfaces and the chemically active state of an electrolyte when it is frozen or when the electrodes are disconnected, making an open circuit.

The analogy between an electrolyte and an insulator links Faraday's 1834 theory of electrochemical action to his 1837 theory of static induction in two ways: it is a guide to the possible meaning of experiments and it is a practical guide to their design. It took nearly two years to articulate the analogy experimentally, and by January 1836 Faraday was ready to review many fundamental experiments in electrostatics. Where chemical theories postulated forces moving particles along until they reached the poles where they could recombine, and where electrostatics saw bodies pushed and pulled by attraction or repulsion between particles of electric fluids on their surfaces, Faraday was beginning to see the observable actions – new chemical products, motions of charged bodies – in terms of a larger system of interactions involving the particles of all the matter in the region. A polarised state is necessary for the production of both a voltaic current and frictional electricity, so there should be an electrostatic tension between the two points at which effects appear. Similarly, since conducting matter has little ability to polarise, electric charges (manifested by a tension or tendency to motion) must have their source in the polarised state of nearby non-conducting matter. If the electrostatic force exhibits a tendency to change of state (or place) it must be developed against an opposed tendency: a resistance or reaction. This would mean that conduction and insulation are only apparently different; both are effects of a process in which polar tensions are repeatedly established between particles and discharged as current electricity. By analogy to phenomena like sound, this process would be 'vibratory'. There would be no true 'electrics' because insulators would be merely very slow conductors.

If every force on a particle of a chemically stable compound is balanced all round, then some of this balancing must extend outside the electrolyte itself (see Figure 5.2a). Conceiving electrostatic phenomena in terms of chemical statics rather than mechanical forces between electrical substances, it is possible to imagine that two charged bodies are related over their whole surface, and to everything else in the region, and not just by Newtonian push–pull forces acting along a line of attraction or repulsion. In December 1835 Faraday was experimenting to see if electrostatic phenomena could be detected in a metal dish on the roof of the Royal Institution, a place where no straight lines of action should reach. He reasoned that if forces always act so as to close the circuits of action, so that every action is met by an equal, opposed action, then forces are essentially polar. Faraday now had to evaluate the radical *conceptual* implications of this analogy. He realised that the analogy involved a thought experiment which exposed a problem in traditional methods of measuring charge. According to fluid theories, charge was a quantity of fluid and this quantity was proportional to the strength of attraction or repulsion between charged bodies. If all charge manifests a polarity or tension between bodies, then two identical charges cannot be *directly* related by the repulsive force which tends to separate them. Their very existence requires the existence of an opposite electrical state elsewhere. So, far from being a primary phenomenon, their mutual repulsion needs explanation in terms of a larger and more primary system of interactions in the space between them. According to traditional electrostatics, charge must be collected or accumulated as a discrete quantity (like a cup of water), in order to be measured. But if, as Faraday now thought, all charges are always physically related, then how does a basic measuring device like an electrometer work?

In the eleventh series of his 'Experimental Researches in Electricity' Faraday argued that *induction* – the power of causing an opposite state – is the first principle of electrical science, and perhaps of all science. He gave the word *charge* a new meaning: it refers to the manifest effects of a state of tension sustained by matter. After consulting Whewell about names in December 1836 he called this matter the *dielectric*. Action across the dielectric must be local, by contiguous particles. Charge appears at interfaces between good and poor conductors. This challenged the familiar picture of charge as something that accumulates on the surface of conductors. Dielectrics are necessary to the existence of charge because they prevent opposed forces from neutralising each other. But what about space empty of matter, or the space between atoms?

We have seen how Faraday approached the macroscopic world of electricity through chemistry rather than mechanics. He naturally

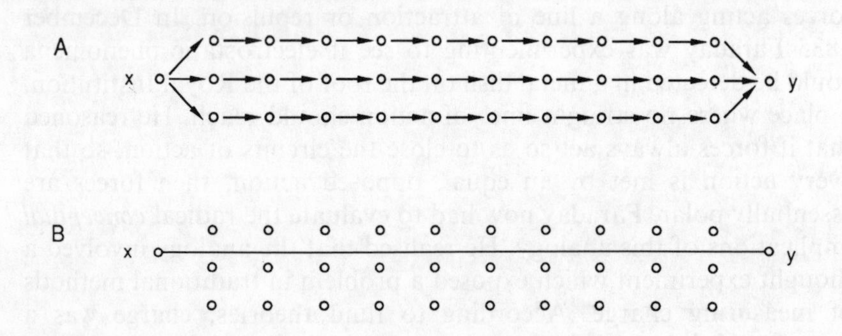

Figure 5.3 Faraday's notion of inductive action. *A* shows induction transmitted between points *x* and *y* by 'contiguous' or neighbouring particles positioned between *x* and *y*. *B* displays action-at-a-distance, which Faraday rejected. Here the action originating at *x* reaches *y* without affecting, or being affected by, whatever intervenes.

assumed that the amount of electricity on 'the surface of the *electric or conductor* must be an almost infinitely small quantity of that belonging to the particles by natural right and giving them their forces of chemical affinity'.[7] Faraday now reasoned from the polar conception of force that electricity acts locally, by 'contiguous' or neighbouring particles, as shown in Figure 5.3a. This ruled out direct actions without affecting the intervening particles, as illustrated in Figure 5.3b. Hence the action must proceed from particle to particle, whatever the distance between them. Although this meant that electrostatic induction needs matter to move between bodies, it left the question of how such action is communicated between the particles of matter unanswered.

5.3 SPECIFIC INDUCTIVE CAPACITY

Faraday was interested in the new experiments suggested by this concept of inductive action by contiguous particles. For example, a charged insulator broken into pieces might exhibit positive and negative charges in the same way that the pieces of a broken magnet do. Moreover, the ability of a substance to sustain tension should characterise it, just as, say, the melting and boiling points do for substances. This led him to predict that the ability or capacity to sustain charge is also specific to each material. The idea surfaces in a sketch for an experiment of 3 November 1835:

> A metal globe in Air; rare air; glass (a body so as to avoid what is called induction); Wax; Oil; Oil turpentine, and other electrics, electrified from the same common source, and then brought in contact with an Electrometer – the divergences observed. If the results are constant for the same body, but vary with different bodies, then a proof that electricity is related to the electric, not to the conductor.[8]

This led to the famous experiments, published in November 1837, in which Faraday showed that the differences in charge (or potential difference) between the outer and inner conductors (labelled *a* and *n* in Figure 5.4) are affected by matter placed in the space *o* between them, and that the effect varies with the type of matter. This shows that insulators affect charge.

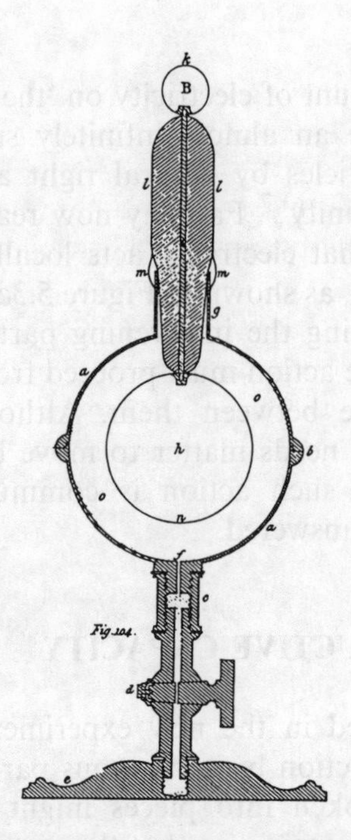

Figure 5.4 Specific electric capacity. The specially designed condenser used by Faraday to prove that the inductive capacity is different for different substances.

Source: *Researches*, vol.1, plate 7.

As usual, other experiments led to other questions. The best vacuum he could produce in his device also maintained a difference in charge between the two surfaces. How could space almost free of matter sustain charge? Still more questions arose from the possibility of charging air: if charge is sustained as tension in insulators (such as air) and not physically held on the surface of conductors, then it should not be possible to give an ‘absolute’ charge of one kind of electricity to matter. In a characteristic move from the invisible realm of atoms to the realm of large objects, he asked:

> If all electricity resides upon the exterior of a globe or mass of conducting matter, then what is the condition of our globe of the earth as a conductor suspended in air?
> Is it electric [charged] within, or not?
> Is it in a state of induction [tension] with respect to the circum-ambient medium?[9]

Atmospheric phenomena suggest that the earth is surrounded by 'an electric; that the electricity resides in the air'.[10] Does the earth therefore carry a certain quantity of electricity, expressible, say, in chemical equivalents? Could this be determined experimentally? He thought not: 'If the globe and its surrounding air were actually electrified with a surplus portion of P[ositive] or N[egative] Electricity, we should not be able to discover it, moving as we do upon its surface and having no other point of comparison than what it affords.'[11]

Perplexed by the problem of an absolute standard for intensity, Faraday was on the verge of concluding that there is no such measure. This paragraph shows him drawing together the threads of an argument that led him to a strange experiment. He built a foil-covered wooden frame so large that he had to move it out of his basement laboratory upstairs to the Royal Institution's lecture theatre. There he 'went . . . and lived in it . . . using lighted candles'.[12] This was the first Faraday cage, now widely used to screen sensitive recording equipment from radio waves. This strange box, large enough to hold several people, originated in an apparently abstract question about comparing positive and negative charges.

5.4 LINES OF ELECTRIC INDUCTION

Faraday is often portrayed as using his experiments on specific capacity to disprove action at a distance. In fact, his search for specific capacity in 1836 was a necessary consequence of other ideas: first, that charge does not exist as a fluid spread over the surface of conductors but results from the tension sustained by nearby insulators; second, that electric forces are *inductive*, that is, they always create opposed (positive and negative) states or effects. He realised that to demonstrate specific capacity does not prove that electrostatic actions cannot cross the apparently empty space of a vacuum or the void between the particles of matter. Faraday did show that the arrangment of charged bodies and the medium around them affects the distribution of electricity. He illustrated this with images of lines of discharge which he had observed through a stroboscope invented by his friend Charles Wheatstone. In Figure 5.5 each line traces the path of a luminous spark or current. Faraday believed that each lit up the path of the line of static tension that must precede any

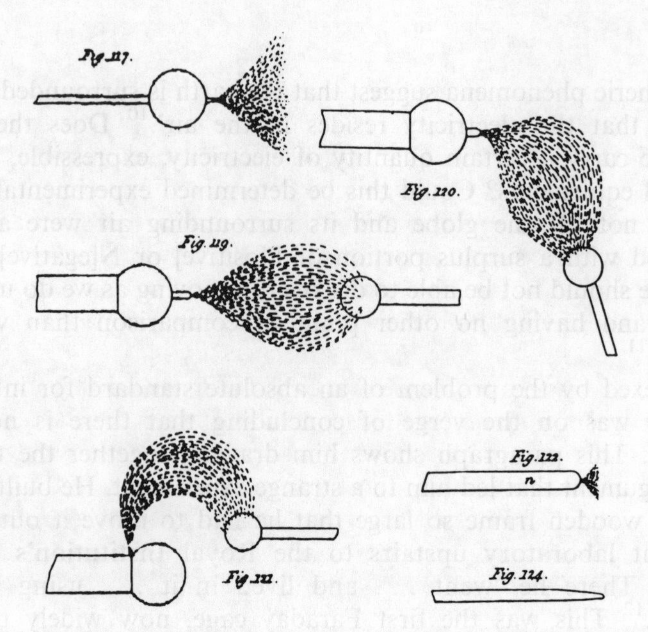

Figure 5.5 Lines of static induction. Engravings of Faraday's sketches showing the appearances of the diffused form of spark, known as the electric brush, when viewed through a stroboscope. Faraday considered each filament or spark-line followed the path of a previously-existing line of tension.

Source: *Researches*, vol.1, plate 8.

discharge of electricity. The pattern of lines changed as the conductors moved, or as new conductors and insulators were introduced, just as iron filings indicate the changes in magnetic lines when magnets are moved or non-magnetic metals are introduced.

An important fact about these lines was that they expand or bulge outwards when there is some distance between the two terminal conductors (see the drawings labelled 119 and 120 in Figure 5.5) but collapse into a narrow line as they are brought close together. Faraday wanted to extend the analogy to magnetism, but he knew that parallel electric currents repel each other magnetically. He developed the image of interacting electric and magnetic forces, first sketched in 1832 (see Figure 4.6), by assuming that electrostatic tension has a hidden magnetic component exerted at right angles to the lines of induction; this would account for the lateral bulging of the lines in Figure 5.5. When the electric tension gives way to current electricity and the lines collapse into a current this 'lateral' force becomes the magnetic field of the current. Inductions and discharges follow each other at a very high frequency. Once again, he seemed close to understanding how electric and magnetic forces might propagate as a wave. Soon afterwards Faraday's progress was blocked. If, as he believed, electricity and magnetism are just different forms of the same force, then magnetism should affect all matter, as do electricity, gravity and heat. Early in 1836 he tried to determine whether metals (other than the known magnetic metals: iron, nickel and cobalt) become magnetic at very low temperatures. In 1838 he made several attempts to detect the 'lateral' or magnetic component of the charged (electrostatic) state. He also tried to detect a magnetic effect analogous to specific electric capacity. None of these experiments produced the expected results. No other materials seemed to affect the action of a magnet at all, although induced currents could be produced in insulators like copper. The restriction of magnetism to just three metals appeared to preclude a general theory of electricity and magnetism.

5.5 MATTER, MAGNETISM AND LIGHT

This obstacle disappeared at the end of 1845 when Faraday showed that magnetism affects all kinds of matter. This encouraged him to incorporate magnetism into his emerging theory of matter and forces. The resulting statement challenged the primary explanatory role that Newton had established for ponderomotive or 'push–pull' forces. We shall see that, according to Faraday, attractive and repulsive forces are not the primary causes of motion that mechanical theory held them to be. Instead, he believed that motion could be explained by electric and magnetic fields. In the mid-1840s another equally radical part of his view was that matter is nothing but the sum of its manifest powers, or, more precisely, no other kinds of matter can be known to human observation. In 1837 Faraday the chemist had challenged established electrical theory. By 1846 Faraday the experimental philosopher openly challenged the most fundamental conceptions of the whole Newtonian tradition.

How did he discover the new magnetic phenomena that gave such a boost to his emerging vision of nature? We saw earlier that, like most of his contemporaries, Faraday believed that each type of force should be converted into others. The starting point was to show that one (such as electricity) can affect another (such as magnetism). In

1823 John Herschel had tried to affect light with an electromagnetic spiral. Since the early 1820s Faraday had tried and failed several times to make static electricity affect light, hoping that optical phenomena would reveal the effect of electricity on matter itself. At the June 1845 meeting of the British Association for the Advancement of Science he met William Thomson who was exploring analogies between mathematical descriptions of different forces. Thomson was then only 21. Their conversation and Thomson's letters encouraged Faraday to return to the search for an electrical effect on light. Typically, early experiments showed nothing. In September he tried magnetic forces instead of static electricity. Faraday had two theories in mind: either magnetism would affect light directly or light would indicate an effect on matter. To increase this possible effect he needed glass with a high refractive index. At hand was the very dense glass he had made in 1829–30 for the Royal Society. He placed a piece of this between the poles of a very powerful electromagnet. Faraday's sketch of its two poles is shown

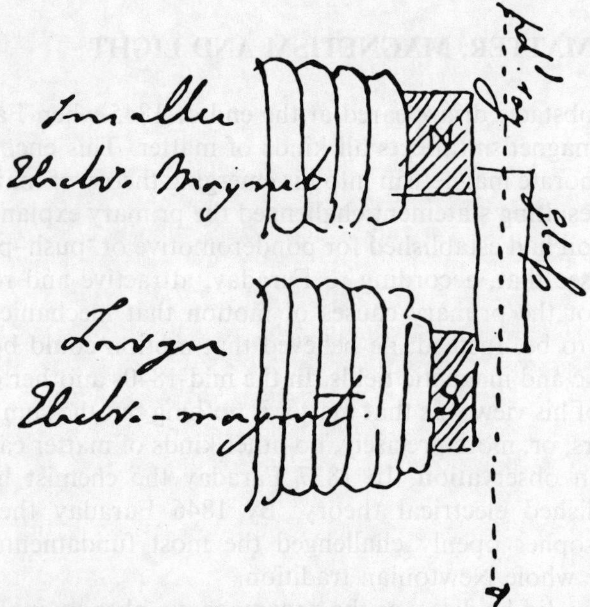

Figure 5.6 The magneto-optic or 'Faraday' effect. With this arrangement Faraday discovered that magnetism affects rays of plane-polarised light passing through dense glass, by rotating the plane of polarisation.

Source: *Diary*, vol.4, p. 264.

in Figure 5.6. Passing a ray of polarised light through this, parallel to the lines of force running between the poles, Faraday found that the magnetism affected the light. How did he make it visible?

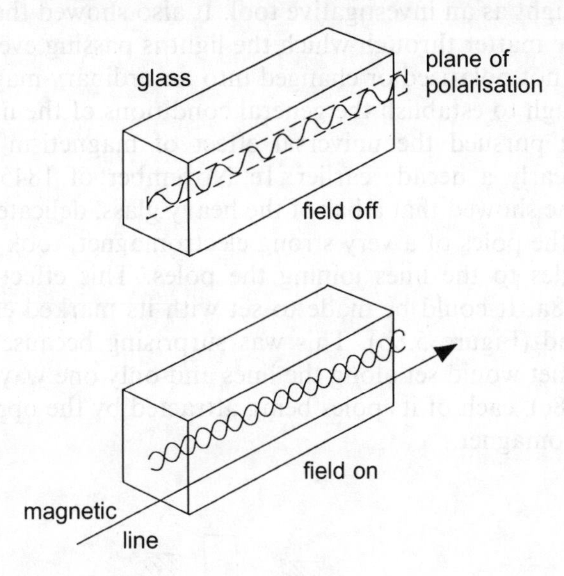

Figure 5.7 Illustration of the 'Faraday effect' showing how the plane of polarisation of the light ray is rotated by the magnetic field.

He found that magnetism changes the orientation of the plane in which the light ray is polarised (see Figure 5.7). This effect, known as the Faraday-effect or magneto-optic effect, existed only if the light-rays passed along the lines of magnetic induction between the poles. Since Faraday thought of light-rays as vibrations, he described the conditions needed for the effect in terms of these lines that filled the space between the magnet poles. Here, for the very first time, he called this space the *field*. It is interesting that he named this, perhaps the most famous of his conceptual discoveries, using an ordinary English word, and without consulting anyone. 'Field' denotes an area of activity (such as a field of battle) or of attention (such as the optical field of a microscope). Faraday always understood the field as a space filled with lines of electric or magnetic force. He thought that the polarised ray illuminated the course of these lines, remarking in his *Diary* that this fact would 'prove exceedingly fertile and of great value in the investigation of . . . natural force'.[13]

5.6 DIAMAGNETISM

The magneto-optic effect vindicated Faraday's belief in the importance of light as an investigative tool. It also showed that magnetism affects the matter through which the light is passing even though the matter is not polarised or changed into an ordinary magnet. Pausing long enough to establish the general conditions of the magneto-optic effect, he pursued the universal effect of magnetism that he had sought nearly a decade earlier. In November of 1845, after many failures, he showed that a bar of the heavy glass, delicately suspended between the poles of a very strong electromagnet, took a position at right angles to the lines joining the poles. This effect is shown in Figure 5.8a. It could be made to set with its marked end (*X*) either way round (Figure 5.8b). This was surprising because an ordinary iron magnet would set along the lines and only one way round (as in Figure 5.8c), each of its poles being attracted by the opposite pole of the electromagnet.

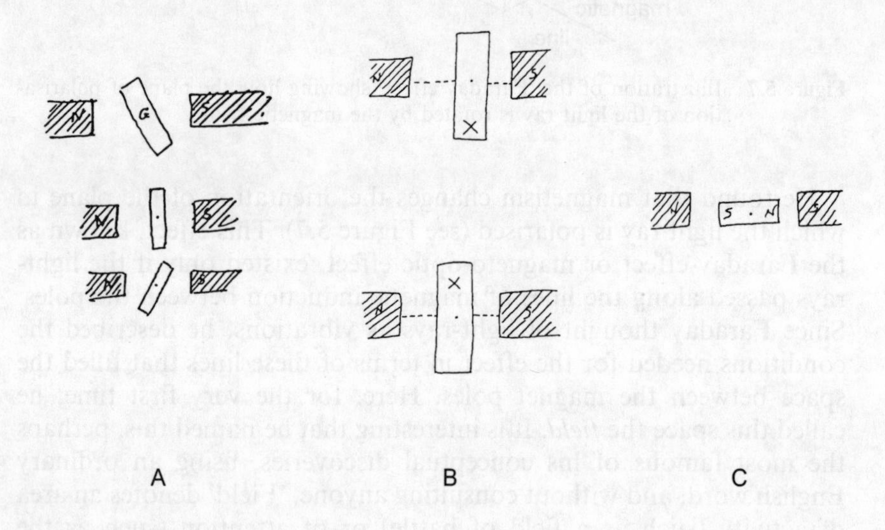

Figure 5.8 Diamagnetism. Figures *A* and *B* show that a diamagnetic bar sets perpendicular to the lines of magnetic force. By contrast a magnetised iron bar sets along the lines, and then with opposite poles adjacent – as in *C*.

Source of A: *Diary*, vol.4, p.313.

This is one of those occasions in the history of science when originality has less to do with observing a new phenomenon than with the ability to appreciate its significance. Other scientists had observed this effect before Faraday, but since it did not appear to involve attraction or repulsion (as expected on the Newtonian theory) they did not pursue the matter. Faraday now showed that the effect is general because bars of wood, paper, metals and many other substances will set across the lines of force. In his *Diary* he noted that 'if a man could be in the Magnetic field, like Mahomet's coffin, he would turn until across the Magnetic line, provided he was not magnetic'.[14] This entry combines the possibility of a fine public demonstration with an important question: if the man is not magnetic, what sort of magnetism would be at work? It was not polar in the way that ordinary magnets are polar, and neither did matter affected in this way acquire its own magnetism. The new 'magnetic' properties depended entirely on the presence of a strong external source of ordinary magnetism. Faraday soon showed that substances can be arranged in order of their sensitivity to magnetism. He was initially surprised to find that gases were not affected, and it was nearly two years before he showed that they were. He asked himself, as he had in 1836, 'What ought a vacuum to do?'[15] Once again, the physical role of empty space was ambiguous. Thin evacuated tubes of fine glass exhibited no effect, suggesting that empty space occupies an intermediate point between ordinary magnetic substances, like iron, and the much larger class manifesting this new effect, such as glass, copper and wood. On the other hand, both magnetic and electric forces act across a vacuum.

At first Faraday called these substances dimagnets, but after consulting Whewell he named them *diamagnets*, by analogy with dielectrics. This term also distinguishes them from ordinary magnets. The key criterion of the magnetic identity of a substance was whether it set 'equatorially' or across the lines of force, which identified it as a diamagnet (as in Figures 5.8a and b) or 'axially', along the lines of force, as do ordinary magnets (as in Figure 5.8c). Faraday then analysed the behaviour of diamagnets more closely. The new magnetic properties existed only while the field was switched on: unlike ordinary magnets, diamagnets cease to affect each other when no strong magnetic field was present. He found something else that impressed him greatly: diamagnets move differently from ordinary magnets. Whereas ordinary magnets are attracted to poles, dia-

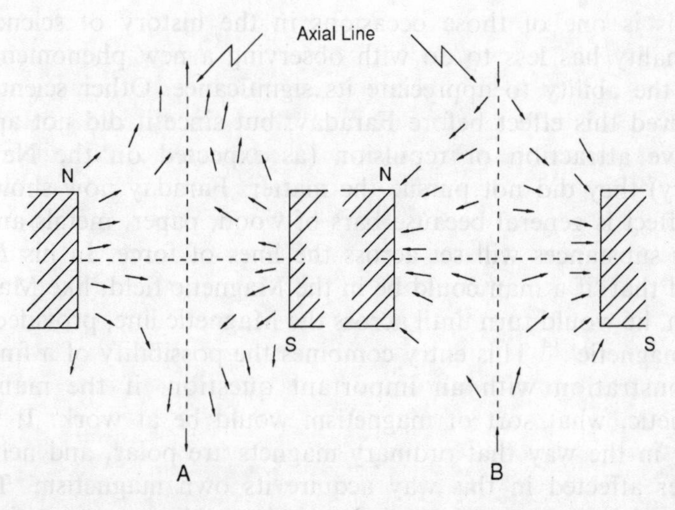

Figure 5.9 The law of diamagnetic action. The contrast between the motion of diamagnets (as in *A*) and those of ordinary magnets (*B*) led Faraday to the general statement that the former move into places of weaker action while the latter move into regions where the magnetism is strongest.

magnets move away from them. The general law of action was that diamagnets move into regions further from the poles, where the magnetism is weakest (as in Figure 5.9a), while magnets move into places near the poles, where the magnetic action is strongest (see Figure 5.9b). Faraday became convinced that these new phenomena were not caused by forces pulling the matter to, or pushing it from, the poles. Faraday's law describes the behaviour of objects in terms of the field, rather than attributing them to some inherent tendency to be attracted or repelled by forces emanating from the poles of magnets. This was a profound change of view.

Why did Faraday find this new picture so appealing? It is reminiscent of his earlier claim that chemical change took place throughout an electrolyte. This made it unnecessary to postulate attractions or repulsions of particles of matter by the two poles. Faraday's chemical approach to physical matters is certainly relevant. Just as important was his conception of a providentially ordered creation. He believed that God has ordered things so that there is no waste of physical action in nature. This global principle would be manifest in the minutest phenomena. Magnets move into regions of most intense action because the force passes through them more easily. Diamagnets move out of regions of intense magnetic action because they conduct it less efficiently than magnets do. Generally, bodies should move so as to reduce the amount of resistance to the passage of force, where the natural motion is the one that realises the most economical distribution and expenditure of force. This simple idea implies a non-Newtonian view of motion: it means that the behaviour of any body in the magnetic field is not due to its having any intrinsic quality such as mass, but is determined by whether it is a better or worse conductor of lines than the surrounding matter.

He now demonstrated the differential conduction idea with fine glass tubes containing solutions of copper sulphate. A solution immersed in a much weaker solution behaved like a diamagnet, but that same solution behaved like a magnet when immersed in a stronger solution. We saw earlier how Faraday had concluded that electrical charge, conductivity and resistance are relative rather than absolute properties. Although this illustration was not a proof, it showed the plausibility of extending the idea of relative conductivities to magnetism. The motions of bodies in the field could be explained by reference to what filled the field (matter and lines of force) rather than by attraction and repulsion. Matter moves so as to offer least resistance to the passage of tension or strain, while lines of force arrange themselves so as to follow paths of least resistance to the passage of inductive action.

5.7 AETHER AND MATTER

This account fitted Faraday's theological ideas about the economy of nature. However, it required empty space to conduct lines of force. This requirement conflicted with traditional ideas about matter and space (space being empty and possessing no properties except extension). Yet there was evidence for such action at a distance; in the best vacuum he could make, magnets and diamagnets were still affected. So space had to transmit magnetic as well as electrostatic action. This called for a new theory about matter and space. He published his speculative ideas on this subject in lectures on matter and force given at the Royal Institution in 1844 and 1846. In the first of these lectures he attacked traditional atomistic conceptions of matter, arguing that matter can be known only by its manifest properties and should therefore be conceptualised in the same way as electricity or any other object of interest.

Emboldened by his discoveries of the magneto-optic effect and diamagnetism, Faraday went even further in a lecture of April 1846 offering his 'Thoughts on Ray-Vibrations'. He argued that electricity and magnetism are transmitted through space as vibrations in lines of force and that other powers may do so too. Earlier he had banished imponderable electric fluids; now the aether – a fluid comprising very minute material particles which was considered by most contemporary scientists to be suffused throught space – followed in their wake. If, as he now argued, aether is just a highly rarefied form of matter, then there was no need to postulate a special substance in which light and other rays travel. Faraday went further still: having dismissed the aether, leaving only vibrations of lines of force, he did away with atoms as well. These, he argued, are only imaginary points. Their inertia – an essential attribute of material atoms since Newton – is due to a 'sluggishness' in the lines of force connecting each point to every other. In other words, the inertia of every atom – its resistance to being set in motion (or having its motion altered) – is like the tendency of knots in a fishing net to resist being moved. This inertia is not a property of the atom (knot) itself but of the physical connections through lines (or strands) to every other part of the field (or net).

It is not surprising that some criticised Faraday for divesting matter of its primary status in the scheme of things. This reflected their allegiance to traditional, well-tried theories of matter and force. Also, Faraday had as yet only sketched a view that took several decades and several brilliant theorists to develop into the theory that eventually replaced the Newtonian mechanical view. Faraday reserved these speculations for lectures because he knew that a great deal more work was needed to make them plausible to others. There were still phenomena that did not fit the theory. The most important was that gases appeared indifferent to magnetism whereas, according to his hypothesis that aether is attenuated matter, gaseous forms of matter should show much more susceptibility to magnetic influence than his vacua did in his experiments. As a result, he did not know how to analyse space.

In October 1847 he learnt that an Italian scientist had shown that magnetism affects flame. This was the clue he needed. Faraday soon repeated the experiment to include streams (or stationary samples) of gases in an atmosphere of carbonic acid gas, in order to observe their tendency to move into or out of the region of most intense action between the poles. The most delicate experiments drew on work on soap films by Joseph Plateau. Faraday introduced soap bubbles of different gases into a magnetic field carefully constructed to eliminate air currents. A few gases, notably oxygen, behaved like an ordinary magnet would, moving into the 'axial line' (see Figure 5.9b). According to his conduction hypothesis these gases would be better conductors of magnetic action than the air displaced by the bubble. Yet oxygen could not be magnetised in the manner of a metal like iron. Faraday realised that he needed three classes of magnetic matter. He created a new class of materials that move like ordinary magnets yet have no polarity or permanent magnetism outside the field. Another consultation with Whewell produced a new name for the former: *paramagnets*. Materials that move out of the field were *diamagnets*, and the metals that can be magnetised and keep their polarity like iron he now called *ferro-magnets*.

Like Faraday's electrochemical and electrostatic terms, these names are still used today. He introduced them partly because he needed to distinguish between diamagnets, like bismuth, and non-ferromagnetic substances, like oxygen. Another reason was his increasing confidence in the hypothesis that magnetic phenomena can be explained in terms of conductivity. A series of elegant and very delicate experiments on crystalline diamagnets helped refine his understanding of the relationship between light, magnetism and matter. He had shown in 1845 that magnetism affects the plane of polarisation of light rays when the light rays and the magnetic lines are parallel. The new experiments of 1848 showed that the optic axis of a crystal – the line along which a ray of light is doubly refracted – is also the axis along which magnetic lines are conducted most readily. In other words, light, as well as magnetism, obeys a fundamental principle of God's economy: bodies move so as to permit the most efficient transfer and expenditure of forces while forces seek paths of least resistance.

5.8 FARADAY'S FIELD THEORY

From 1850 Faraday began to publish a comprehensive theory of magnetism. This began with small-scale laboratory phenomena but he soon extended it to include global, atmospheric phenomena. He argued that all matter transmits magnetism but whereas some things merely conduct, others, such as iron, acquire and retain magnetic power. This theory required that diamagnets are worse conductors than empty space while paramagnets are better conductors than space, though not as efficient as ferro-magnets. This resolved the recurring problem of relative conductivity. In 1835 Faraday had asked whether the measure of electrostatic charge could be based on any absolute standard. Placing space between para- and diamagnetics in his tri-partite distinction provided an answer: make empty space the middle-point. The behaviour of lines of induction (as shown in patterns of iron filings) gave partial support to this idea. Faraday had known since he began work on electromagnetism in the 1820s that soft iron increases or intensifies the magnetism of a current. One way of describing this was to say that iron concentrates the lines of force so that there are more lines of action packed into the space occupied by a magnet than elsewhere. Filing patterns show that lines converge or concentrate as they enter the iron and diverge as they leave it (see Figure 5.10). The electromagnet – a wire wound around a soft iron

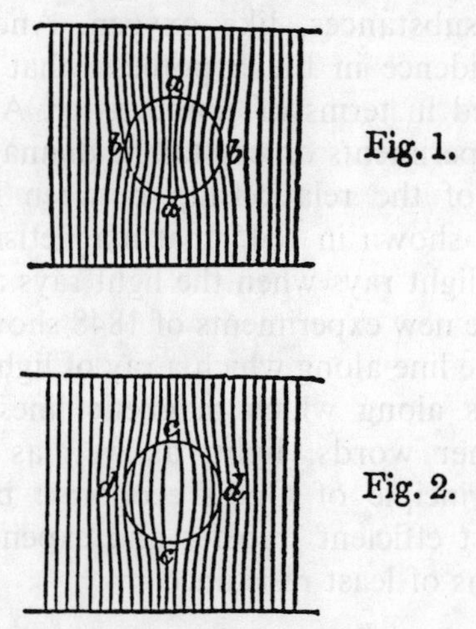

Figure 5.10 Convergence and divergence of lines of induction. Moving from top to bottom in the upper drawing, the lines concentrate or converge as they enter the magnet and diverge as they leave it. In the lower drawing, the lines diverge as they enter the diamagnet and converge as they leave it.

Source: *Researches*, vol.3, p.204.

core – was a well known practical application of this concentrating action. Faraday's conduction hypothesis explained why lines should behave this way. However, he was never able to observe the diverging effect that a diamagnet should have on lines in, say, an atmosphere of a better conductor such as oxygen.

There was also a metaphysical difficulty. Most of his contemporaries defined space in terms of geometry (as a place where physical events may occur) but not physically (it could not participate in physical processes). Faraday's theory required that space participates: it was required to conduct better than diamagnets do. His *Diary* shows that he treated vacua in just this way; he examined the behaviour of evacuated glass tubes, including the results in a long list of experimental samples. In his experiments he always dealt with a vacuum, which would contain highly rarefied matter as space wholly devoid of matter was but an imaginary ideal. Void space had no special, metaphysical status for Faraday.

Faraday's mature concept of the field therefore makes a theoretical necessity out of an experimental virtue. Faraday came to think of space as the field, a place that enables things and observers to interact. The so-called aether is equivalent to space as he now understood it. Both had the capacity to transmit the powers of matter. Faraday classified substances according to their powers, so that space possessed fewer powers than say, a diamagnetic like bismuth, which in turn had fewer powers than a ferromagnetic like iron. Material powers such as magnetism or gravity would be transmitted through apparently empty space by lines and, as he had suggested in 1846, light would travel as vibrations in such lines.

By 1850 he had extended his theory beyond the local phenomena of the laboratory, to explain the effect of the condition of the earth's atmosphere on global variations in the earth's magnetic field. Gravitation continued to elude him. In experiments made in 1849 he failed to detect the predicted relationship between gravity and electricity. There was still much to do. He was nearly 60 years old and realised that the experimental facts already published in the many series of experimental researches were not persuasive enough. Could the evidence be made more conclusive? Between 1850 and 1855 he concentrated on arguments for the reality of fields. This meant marshalling old evidence rather than looking for new empirical clues. Nevertheless, he continued to seek experimental confirmation of the conversion of electricity and gravity and in 1856 he also began new research on solutions of finely-powdered metals. This work on suspensions of gold is widely regarded as the beginning of colloidal chemistry.

However, Faraday's main concern in the last decade of active scientific work was to determine whether the theory of lines of force was merely a useful representation, or whether it truly described the world. He now had a greater experimental knowledge of electric and magnetic phenomena than any of his peers. He used this knowledge to select complementary aspects of these phenomena to develop a new, dynamic image. For example, where in 1831 he had seen that electricity, magnetism and motion are exerted at right angles to each other (Figure 4.6) he could now state this relationship in more general and more rigorous terms. As he wrote in 1852,

> Let two rings, in planes at right angles to each other, represent [the electric and magnetic forces] . . . as in [Figure 5.11]. If a current of electricity be sent round the ring E in the direction marked, then lines of magnetic force will be produced, correspondent to the polarity indicated by a supposed magnetic needle placed at NS, or in any other part of the ring M.[16]

The interlocking rings in Figure 5.11 can represent either the current induced by a magnetic field or the field produced by a current. Both the current and the magnetic line can be described as a line or axis of power that is polar (i.e., having opposite effects or properties in

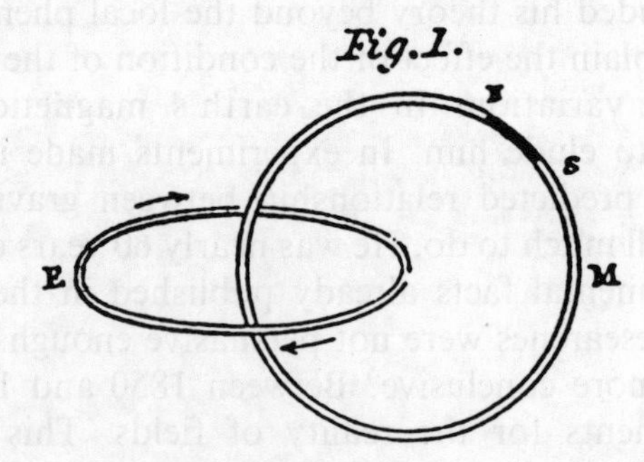

Figure 5.11 Interlocking rings of electricity and magnetism.

Source: *Researches*, vol.3, plate 4.

opposite directions). Moreover, as he had realised in 1837 when experimenting on electrostatic lines, these tend to bulge laterally when there is an electric charge but collapse and elongate when this is discharged as a current (see Figure 5.5). Lines of magnetic induction around a current tend to repel each other but attract those of an oppositely directed current. Faraday now assembled these complementary aspects of the polarity of electricity and magnetism into a model of an electromagnetic wave in which, as the electric component increases, so a magnetic component perpendicular to it decreases, and as the magnetic increases, the electric decreases. Thus he could at last delineate these two forces as inter-convertible forms of a single power.

This view was now clear enough to publish. Although the papers containing his most important arguments for the physical existence of lines of force continue the paragraph numbering of his series of 'Experimental Researches in Electricity', he published them as independent articles in a different journal. Faraday probably felt that, however plausible his model of electromagnetic vibrations was, this could not prove that lines of force really exist, and neither could he totally disprove the alternative theory. This alternative, which was widely supported on the Continent by Ampère, Weber and others, postulated forces that act at a distance: thus one body may influence another by a force that can travel between them instantanously and without affecting, or being affected by, any point in between. Faraday's dislike of the idea of instantaneous, unmediated action emerged from other beliefs, on which he now drew in his arguments for the reality of lines of force. Of these, the most important are the polar or dual nature of all forces (except, apparently, gravity), the conduction hypothesis, the indestructibility or conservation of force, and Faraday's disregard for quantities defined or measured in relation to distance, or 'mere space'.

He combined polarity and conduction to explain why a magnetised steel ring exhibits little or no external field. Cut the ring, however, and field-lines immediately appear around the cut. This is easily shown by iron filing patterns, which take the forms shown in Figure 5.12. That most of the power imparted to the magnetised ring seems to stay inside it showed that iron conducted the magnetism very much better than the surrounding air (or *medium*, as Faraday now called it). Just as when an electric circuit is broken, the force passes as a spark through a poorer conductor filling the gap. The lines diverge

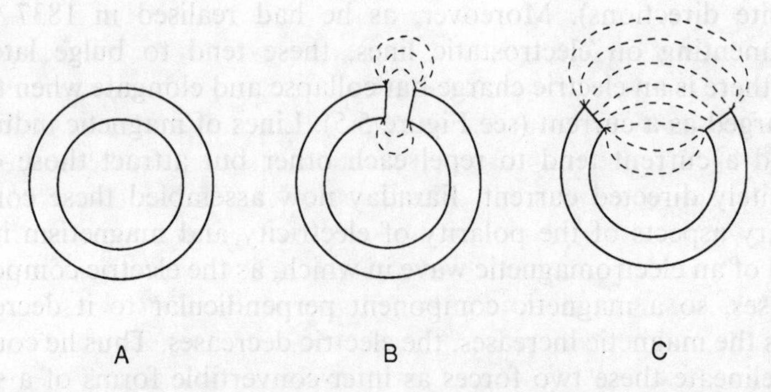

Figure 5.12 The effect of cutting a steel ring magnet. Figure *A* shows a closed magnetised ring. When cut (as in *B*) an external field appears. As the ring is opened out into a horseshoe, *C*, the external field extends considerably.

as they leave the magnet (Figure 5.10) or the conducting wire (Figure 5.5), and converge again as they re-enter. This was an instance of the duality of forces, namely that every polar force exhibits opposite effects in opposite directions (for example, reversing the direction of a current reverses the polarity of its magnetic field; reversing the motion of a wire moving near a magnet reverses the direction of the current induced in it, and so on). This showed that it is impossible to have force of just one kind. In other words, it would be impossible for an action to emanate into space without an equal and opposed reaction.

In 1837 Faraday had argued that with static induction there can be no 'absolute' charge, that is, no free electricity of one kind. He now argued that there are no isolated or free magnetic poles. Faraday could infer from the effects shown in Figure 5.12 that poles exist only when (and because of) the reactive medium introduced by cutting the iron ring. For ordinary bar and horseshoe magnets this meant that he considered the 'outer medium as *essential* to the magnet'.[17] No magnet could exist without it because the medium 'relates the external polarities to each other by curved lines of power'.[18] This made the conducting and reactive properties of media, including space, essential. As we saw earlier, this line of argument originated in Faraday's concept of electrochemical polarity, applied in his analogy between an electrolyte and a charged condenser (see Figure 5.2).

Since Faraday conceived force as a polar, conserved quantity of action, such evidence argued the necessity of admitting a physical role for space. If no ordinary matter is present, space must provide the necessary reaction or resistance, without which force could not exist as a 'polar relation'. This was the physical and metaphysical context for Faraday's main piece of evidence, the induction of currents by moving wires. In series 28 of his *Researches* he developed the insight of his first series, that there is a definite relationship between the quantity and intensity of the induced current and (respectively) the number of lines of induction, and the rate at which they are cut by the wire. This work was done with devices like that shown in Figure 5.13. These were specially constructed to allow Faraday to collect and measure definite quantities of induced current. He wanted to show that the current is converted from magnetic power that exists in the space around the magnet. He argued, against the action-at-a-distance view, that the induced current is not created when the wire is brought near the magnet. He could not accept the idea that a quantity of power can come into (and go out of) existence in this way, because only God could create or destroy a conserved quantity like force. Second, current is induced when a wire moves across the lines, or relative to the magnet-poles. But change of position is not a physical cause, because void space could not generate a physical relation between magnet and wire.

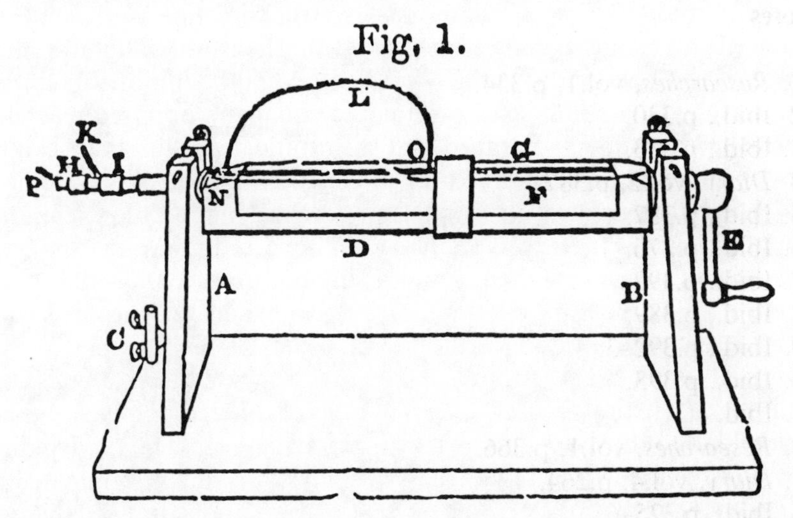

Figure 5.13 One of the devices Faraday used to demonstrate a quantitative relationship between induced current and the cutting of the lines of force.

Source: *Researches*, vol.3, p.333.

Faraday would not accept, as did James Prescott Joule, William Thomson and others, that the induced current is a converted form of the mechanical force exerted to move the conducting loop, L, in Figure 5.13. He was unable to accept the concept of potential energy as a quantity that increases as two attracting bodies separate, so compensating for the decreasing intensity of the attractive force between them. His conviction that mechanical force exerted in turning the crank of the generator in Figure 5.13 does not convert to electricity made him confident that electromagnetic induction demonstrated the existence of lines of power conducted through the space around the magnet.

In this chapter we have seen how Faraday learnt his science in a practical world of ordinary objects and substances defined by qualities such as their mass, density, position or reactive properties. These things were affected by chemical and mechanical forces, defined in terms of the amount of change they produced. Faraday introduced new ways of manipulating chemical, electrical and magnetic forces, and a new way looking at this world in terms of continuous fields of undulating lines of force.

Notes

1. *Researches*, vol.1, p.334.
2. Ibid., p.330.
3. Ibid., p.333.
4. *Diary*, vol.2, p.387.
5. Ibid., p.387.
6. Ibid., p.276.
7. Ibid., p.392.
8. Ibid., p.389.
9. Ibid., p.392–3.
10. Ibid., p.393.
11. Ibid.
12. *Researches*, vol.1, p.366.
13. *Diary*, vol.4, p.264.
14. Ibid., p.325–6.
15. Ibid., p.351.
16. *Researches*, vol.3, p.418.
17. Ibid., p.425.
18. Ibid.

6 Faraday's Influence

Faraday's work has had an impact on many different fields of activity. From electrical engineering, through electrotherapy screening, to the upper reaches of theoretical physics, Faraday's name is to be found. In this final chapter we examine three of the many areas affected by him and his work. First, we look at the effect of his scientific research on certain areas of science and engineering; second, the public image of Faraday; and, finally, we point to some of the ways in which this study of Faraday should influence our understanding of science.

6.1 SOME INFLUENCES ON SCIENCE AND ENGINEERING

Field theory

Many of Faraday's discoveries were incorporated into the edifice of science in his own lifetime. Electromagnetic rotation, electromagnetic induction, diamagnetism and the laws of electrochemistry, to name but four, were soon generally recognised by contemporaries as significant contributions to positive knowledge. While these discoveries were highly important contributions to mid-nineteenth-century science, we misunderstand many aspects of his work if we concentrate specifically on his discoveries. In the two preceding chapters we presented his discoveries in the context of Faraday's developing understanding of the world over a period of half a century. Faraday's odyssey began with his training in chemistry and ended with his new natural philosophy based on field theory. This speculative thread running through Faraday's thought exerted a crucially-important influence on the physics of the late nineteenth and early twentieth centuries. However, unlike his positive discoveries, aspects of his field theory were both challenged and also extended in very different directions by others. Modern field theory is based on Faraday's mature insights as developed by other scientists, particularly William Thomson and James Clerk Maxwell.

In the winter of 1855–6 Maxwell, then in his mid-twenties and a Fellow of Trinity College, Cambridge, wrote a paper entitled 'On Faraday's Lines of Force'. As the title indicates Maxwell took one of Faraday's speculative ideas as his starting point. However, Maxwell possessed a formal training in physics and mathematics and was able to analyse lines of force in ways that were not acceptable or accessible to Faraday. The major thrust of Maxwell's project was to express Faraday's lines in a mathematical language. He was also concerned that this could be a sterile exercise unless the mathematical formulae were firmly anchored in physical reality. The anchor was provided by experiment and physical analogy; Maxwell drew the analogy between the behaviour of electric and magnetic lines, on the one hand, and the motion of an hypothetical frictionless fluid on the other. He was not claiming that Faraday's lines *were* the motions of a frictionless fluid, but rather that he could gain insight and understanding of these lines by articulating this mechanical analogy in mathematical terms. This strategy moved discussion of Faraday's lines in a new and exciting direction but one which Faraday could not fully appreciate.

In the late 1850s Maxwell carefully studied Faraday's papers on field theory (see section 5.8) and also corresponded with him. Like Faraday, Maxwell recognised the problems inherent in the action-at-a-distance theories of electromagnetism accepted by Ampère, Weber and others, and he saw the possibilities of developing the mathematical and analogical understanding of field theory. In correspondence he also pointed out that Faraday had erred in his analysis of gravitation and especially in claiming that the gravitational force acts in straight lines and is therefore unaffected by the medium. On the contrary, Maxwell argued, gravitational lines of force are modified by the distribution of matter and since gravity forms a field it can be analysed in a similar way to the electromagnetic field. Thus fields have a far wider applicability than Faraday had recognised and the gravitational force, which Faraday considered to be anomalous, was brought into closer conceptual alliance with the other physical forces.

Maxwell's main engagement with field theory was published in 1861–2, under the title 'On Physical Lines of Force'. As in the earlier paper his analysis was grounded in both mathematics and mechanistic analogy. Maxwell conceived Faraday's electromagnetic medium filling all space as a fluid which forms small molecular vortices. These vortices are in motion and Maxwell treated them as a mechanical engineer would, since he introduced 'idle wheels' to enable adjacent wheel-like vortices to revolve in the same direction. If this analogical method sounds laboured it nevertheless yielded some impressive results. Maxwell was able to develop parallel sets of equations – 'Maxwell's equations' – linking the electromagnetic and mechanical systems. For example, the angular velocities of his vortices correspond to the magnetic field intensity. Most importantly he showed that light is the transverse vibration of the electromagnetic medium and this vibration travels at the speed of light. Faraday's speculation in his 1846 'Thoughts on Ray-Vibrations' paper (see section 5.5) was now firmly grounded in mathematical theory!

Albert Einstein's development of the special theory of relativity in 1905 arose from a critique of Maxwell's interpretation of Faraday's law of electromagnetic induction. He identified an apparent contradiction in the Maxwellian analysis between two cases of electrical induction: first, when the conductor is stationary and the magnet moving and, second, when the magnet is stationary and the conductor is moving. Although Einstein criticised this aspect of Maxwell's theory, he saw relativity theory as 'the direct outcome and, in a sense, the natural completion of the work of Faraday, Maxwell, and [Hendrik Antoon] Lorentz'.[1] Moreover, he conceived Faraday as initiating electromagnetic field theory which he portrayed as the crucial break with the earlier mechanistic view of nature. This 'great change', wrote Einstein, 'will be associated for all time with the names of Faraday, Maxwell, and [Heinrich] Hertz'.[2] Although historians sometimes portray Einstein as breaking with nineteenth-century physics, it is important to recognise this continuity.

Einstein was not alone in attributing to Faraday one of the most profound theoretical innovations in the history of physics. Let one further example stand for many: in 1907 a young Cambridge experimental physicist and philosopher named Norman Campbell wrote *Modern Electrical Theory*, a book which summarised and integrated the contemporary understanding of electricity. Campbell, who later held a Fellowship at the University of Leeds, attributed to Faraday the '"medium" interpretation of physical forces' and then proceeded to claim that Faraday's researches 'are the foundation of the science [electricity] which is the subject of this book'.[3]

Electrical technology

As an example of Faraday's influence on technology we will cite the electric telegraph. Faraday played only a minor role in its invention, in that he advised one of the inventors, William Fothergill Cooke, that his ideas were sound. Only when Cooke met Charles Wheatstone, Professor of Experimental Philosophy at King's College, London, and a close friend of Faraday's, were they able to invent and patent a practical electric telegraph. Its first use was in 1838 from Paddington to West Drayton, along the newly-built Great Western Railway. Thereafter Britain was rapidly wired for telegraphic transmission along the tracks of the burgeoning railway system, as also happened on the Continent.

There were no problems until 1848 when a special continuous underground telegraph cable was laid from Frankfurt (where the German Diet – parliament – was meeting) to Berlin, so that the Prussian government could know quickly what was being decided at the Diet. Until this time telegraphic messages had only travelled over relatively short distances. The engineer in charge of the cable, Werner Siemens, observed that signals underwent what became termed 'telegraphic retardation' in that a sharp signal sent from Frankfurt was 'blurred' when it reached Berlin. This phenomenon only excited sustained concern when it appeared in the first cross-Channel cables laid in the early 1850s linking Britain with France and Belgium. At this time a project to lay a further cable from Ireland to Newfoundland was also commenced but, in view of the greater distances, the problem of blurring became pressing. Approaches based on action-at-a-distance theories of electricity generated mathematics so complex that they failed to provide a solution to the problem. William Thomson (later Lord Kelvin) was asked by the Atlantic Telegraph Company to investigate. Instead of pursuing action-at-a-distance theories, he approached the problem using Faraday's theory of the electromagnetic field (see section 5.8). Thomson's solution involved the resistivity of the copper wire and the electrical capacity created across the gutta-percha insulator which encompassed the wire. Field theory highlighted the importance of the diameters of the wire and the insulator. By increasing these newly-appreciated parameters the blurring could be reduced. However, the cable could not be made very thick since the cost of production and the difficulties of laying the wire on the ocean bed increased enormously. The first transatlantic cable was laid in 1858, but it broke soon after. A second cable was laid in 1866, again under Thomson's direction.

At the beginning of Faraday's life messages could travel little faster than in ancient times. By the time of his death, messages could be transmitted to the New World in a matter of minutes. The theory drawn from Faraday's experiments had played a major role in bringing the first electrical information technology into existence. By contrast, the early development of electromagnetic technologies had been apparatus-based. The design and construction of motors and generators was fairly advanced before theory-guided problem solving and the introduction of design improvements became important. Faraday's rotation motor showed that continuous mechanical effect (motion) could be derived from chemical forces via their conversion into a current whose field interacted with that of the magnet (see section 4.2). Faraday did not work out the practical applications although he recognised the conditions necessary for the interaction of electricity and magnetism to produce motion: they are summarised in Figure 4.6.

During the 1820s many others, such as James Marsh, Gerritt Moll, William Sturgeon, Joseph Henry and Francis Watkins, sought to increase the power of electromagnets, both for demonstrations and for lifting weights. They constructed the earliest working motors and dynamos, principally by the empirical development in the design of electromagnets and devices like Faraday's rotation motor. Numerous such innovations appeared throughout the 1830s. In 1838 Francis Watkins, an instrument maker, claimed that Henry's 1831 device both contained the 'first hint' and was the 'first contrivance wherein electro-magnetic power is made to produce continuous motion'.[4] These claims could equally be applied to Faraday's electromagnetic rotation motor dating from ten years earlier. However, since Henry's design employed an armature operated by the motion of the electromagnet, it was a significant step towards the modern direct-current motor. In 1833 Rev. William Ritchie, the Professor of Natural Philosophy at the Royal Institution, combined Faraday's and Henry's insights in a motor in which the electromagnet rotated continuously. By the end of the decade patents were being sought for motors capable of performing useful work, such as Uriah Clarke's electric locomotive which was illustrated in Sturgeon's *Annals of Electricity* in 1840.

6.2 USES OF FARADAY

In the period since his death Faraday's name and sometimes his portrait have been used for many purposes. His genteel, intelligent face is well known to the public and has graced numerous advertisements including those by electricity generating companies, the makers of electrical goods and even the purveyors of Oxo beef cubes (see Plate 8). His name has been given to avenues, roads and streets, to houses and halls, to prizes and to a professional scientific society. Moreover, the *Oxford English Dictionary* records the Farad, Faradaic, Faradism (or Faradaism), Faradization and Faradize, although only the first of these – a unit of electrical capacity – gained wide usage. Few scientists have become so well known to the general public.

In part the public awareness of Faraday has arisen from campaigns organised by the electrical industry which has been concerned to persuade people to light their homes and cook their meals by electricity rather than by gas. After the First World War the electrification of Britain required a large amount of capital to be raised to build power stations and develop the National Grid. Electrification is a complex issue but one in which the image of Faraday has played a significant role. The centenary of Faraday's discovery of electromagnetic induction was celebrated in 1931 in grand style by the Royal Institution, the Institution of Electrical Engineers and the Royal Society. The events included an exhibition occupying the entire Albert Hall and a commemorative meeting at the Queen's Hall (where the Henry Wood Promenade Concerts were held until it was destroyed in the Second World War). Speakers at the Queen's Hall included Ramsay MacDonald (the Prime Minister and a relative of Lord Kelvin), Ernest Rutherford and William Bragg (holder of Faraday's chair at the Royal Institution). Faraday was amply praised in words and music, the concert being conducted by Henry Wood. There was also an exhibition at the Kingsway Hall (now destroyed) demonstrating the influence of electrical engineering on modern life. These festivities were funded by the Institution of Electrical Engineers at a cost of £10000. Behind the celebrations was the concern among electrical engineers to assure the public of the impeccable scientific pedigree (and thus respectability) of their profession and to help encourage the electrification of Britain. Both of these aims were fulfilled.

Plate 8 Faraday at the hands of the advertiser

Selected aspects of Faraday's biography have also proved useful, although we should note that they have been used for different and even incompatible purposes. Many writers have been particularly impressed by Faraday's transformation from the son of a poor blacksmith to the famous, successful British scientist. This is not quite a rags-to-riches story, since Faraday did not seek wealth, but it has been used to show the efficacy of self-help. Such stories were particularly attractive to many Victorians and their message continues to appeal in political circles. Thus, for example, in a television interview in 1987 Margaret Thatcher, then British Prime Minister, explained why Faraday was her hero and that she had placed a bust of Faraday in the hall of 10 Downing Street. She was impressed not only by Faraday's scientific achievements, on which so much modern electrical technology is based, but also by his ability to transcend his humble origins despite (if not because of) the lack of a formal education. A very different point emerged in a widely-reported Dimbleby Lecture televised in April 1988 in which Lord Porter, then President of the Royal Society, used the example of Faraday to argue the importance of pure research and the shortsightedness of government policy which emphasised short-term technological returns.

The image of Faraday has also served in discussions of the relationship between science and religion. As discussed in section 2.2, Faraday was not only an eminent scientist but also a devout Christian. This combination proved highly congenial to many mid-Victorians who witnessed an increasing opposition of the two, particularly in the area of evolutionary biology and the progressive secularisation of science. Faraday was not only a devout Christian but a man who lived by a literal interpretation of the Bible. Since his death many writers have turned to Faraday to argue not only that revealed religion is compatible with modern science but also that Faraday exemplifies the Christian as scientist. For example, the concluding chapter of a recent book dealing with the relation between science and religion is devoted to Faraday as providing an exemplary instance of this relationship. There are a number of problems with this argument, not least Faraday's adherence to Sandemanianism, a sect that set itself apart from all other churches. Thus, one can but question the legitimacy of, say, an Anglican using this argument when Faraday would not have set foot in an Anglican church. Although there are many problems in using Faraday in this way, he has exerted and will continue to exert a fascination for those who are seeking creative ways to interrelate science and faith. Despite his opposition to spiritualism – see section 2.1 – even spiritualists have invoked his name in support of their cause!

Scientists, too, have often appealed to Faraday's name to legitimate their own approaches to science. Thus Norman Campbell not only attributed field theory to Faraday but also saw himself as the heir to Faraday the experimentalist. He called on contemporary physicists to return to the precise experimental techniques used so effectively by Faraday, and he criticised the abstract mathematical theorising that had come to dominate Cambridge physics by the early twentieth century, and particularly the work of Joseph Larmor, the Professor of Physics. However, Larmor conceived his research, and particularly his programme to build models of the electromagnetic ether, as forming part of a much longer British tradition which had begun with Faraday and continued through Thomson and Maxwell to himself. Thus both experimentalist (Campbell) and theoretician (Larmor) located themselves, and the correct method for pursuing physics, in respect to Faraday.

Many other examples could be cited. However, it should be clear that the image of Faraday has been and remains a potent resource in many different contexts, from selling electricity to justifying both the funding and the underfunding of academic science.

6.3 EPILOGUE

In this book we have departed from the widely-held but narrowly-conceived view of Faraday as a discoverer, first and foremost. This is not to deny that he made some of the most important experimental and theoretical innovations in nineteenth-century science. However, excessive concentration on discovery distorts our understanding of both scientists and the activity of science. In conclusion we wish to draw attention to three aspects of Faraday's work that we have discussed in the preceding chapters: first, the role of general metaphysical, especially religious, assumptions; second, his experimental method, which was complex and pluralistic; finally, his role in increasing the public's understanding of science.

Scientists are often portrayed as open-minded and yet responsive only to hard, empirical evidence gleaned from laboratory experiments. The example of Faraday argues against this view. As discussed in Chapters 1 and 2, Faraday was a member of the Sandemanians who accepted a highly literal interpretation of the Bible. His Christianity can be related to many aspects of his career. For example, as noted in section 3.3, in his dealings with the Royal Institution, the Admiralty and other organisations, he was impelled by a profound sense of duty. Theological themes also played a major role in directing his scientific research. In pursuing science he was driven by his desire to understand the matter, forces and laws that God had wrought at the Creation. He firmly believed that God had created an economical system in which both matter and force were conserved and could not be destroyed. Hence, in his research on the identities of electricities (section 4.4) he was impelled by his belief in the divinely-ordained unity underlying the different sources of electricity and the effects produced. Likewise, his work on diamagnetism (section 5.6) and field theory (section 5.8) was underpinned by his commitment to nature as a conserved, economical system.

To say that Faraday held these metaphysical beliefs is not to claim that he disregarded empirical evidence. We have seen that he took observational and experimental evidence very seriously. However, it does mean that his research was directed by goals and values outside his laboratory. He was strongly predisposed to reject theories that were inimical to him, such as atomism and any that implied the destruction of force, and to develop theories congenial to his metaphysics, such as field theory. His 40-year odyssey can be understood broadly as his increasing dissatisfaction with Newtonian matter theory and his progressive articulation of field theory. Throughout Chapters 4 and 5 we paid attention to the succession of problems Faraday encountered and to how his solutions generated new questions. Viewing Faraday's work in this way places his empirically-based discoveries in a new light.

A second aspect of his work that we have emphasised is the range of practical skills that he brought to his science. His training in chemistry taught him how to construct apparatus and manipulate substances with great skill. Faraday possessed the rare ability to interact with nature with great sensitivity. Rather than prising secrets from nature, he and nature collaborated. In this collaboration he was greatly aware of the role and limitations of his instruments, experimental techniques and his own prejudices. For Faraday experimentation was a highly reflective and a very personal activity; he could not have worked in a modern research group. Nowadays scientists are usually characterised as either theoreticians or experimentalists. While Faraday was precluded from the higher realms of mid-nineteenth-century mathematical physics he combined the theoretical and the experimental very effectively. The modern distinction is not helpful since, in his research, theory and experiment merged. Head and hand were closely coordinated. As the American philosopher Charles S. Peirce remarked, 'Faraday had the greatest power of drawing ideas straight out of his experiments and making his physical apparatus do his thinking, so that experimentation and inference were not two proceedings, but one.'[5]

Faraday displayed other, related characteristics. Although certain themes recur in his research, such as the conservation of force, he displayed great versatility, utilising many different resources and methods. While he manipulated matter principally through its interaction with electricity and magnetism, he also used light in the 1830s and 1840s as another form of probe. This led to his discovery of the magneto-optical effect in 1845 (see section 5.5). Likewise in his discovery of electromagnetic induction he tried connecting the detector before making the battery connection. He also grasped the possibility of magnifying the effect by using a soft iron torus (see section 4.3). Another aspect of his versatility was his recognition that although theories are crucial for their suggestiveness and for helping to direct experiment, they are always provisional. The scientist should treat them as guides but not become attached to them, otherwise they become a source of prejudice and distortion. As we saw in section 4.5 Faraday sought to free the language of science from its contemporary theoretical baggage. His ability to use theories creatively and critically while honestly seeking evidence both for and against them is characteristic of his science.

Third, in our introduction we portrayed Faraday as a communicator of science. Public science lectures achieved popularity by the early eighteenth century, the demand for them increasing dramatically in the early nineteenth. However, Faraday occupies a crucial position in the history of science lecturing. One factor contributing to his success was that he devoted much time and effort, including preparation time, to his lectures. Unlike many of his peers he placed great value on disseminating scientific knowledge. Faraday wanted to spread the fruits of recent research to a wider audience. There was therefore considerable continuity between his roles as a researcher and as a communicator. Since Faraday's death science has become increasingly specialised and professionalised. Research has become separated from teaching. Moreover, for most scientists teaching is seen principally as training students in either research or other professional skills. As a result of the separation of these roles, which may be further institutionalised in the near future, the scientific community has become isolated from the public.

In the 1980s, when research funds were significantly reduced, scientists began to notice the lack of public support. To encourage them to take the public's understanding of science seriously, the Royal Society established a special award, appropriately called the Faraday Award. Such palliatives barely touch the roots of a deep cultural problem; there is a need for scientists to redefine their roles in respect to society. Faraday gave public lectures in order to 'facilitate our object of attracting the world, and making ourselves with science attractive to it'.[6] Does this rationale remain valid 200 years after his birth?

Notes

1. *Nature*, 107 (1921), p.504.
2. A. Einstein, *Ideas and Opinions* (New York: Dell, 1973), p.262.
3. N.R. Campbell, *Modern Electrical Theory* (Cambridge, 1907), p.4.
4. F. Watkins, 'On electro-magnetic motive machines', *Philosophical Magazine*, 12 (1838), p.190.
5. P.P. Wiener (ed), *Charles S. Peirce: Selected Writings* (New York: Dover, 1966), p.272.
6. Jones, vol.1, p.392.

Further Reading

The most recent biographical studies of Faraday are:

L. Pearce Williams, *Michael Faraday: A Biography* (London: Chapman & Hall, 1965).

David Gooding and Frank A. J. L. James (eds), *Faraday Rediscovered: Essays on the Life and Work of Michael Faraday, 1791–1867* (London: Macmillan, 1985; reprinted in paperback, London: Macmillan, 1989 and New York: American Institute of Physics, 1989).

Geoffrey Cantor, *Michael Faraday, Sandemanian and Scientist: A Study of Science and Religion in the Nineteenth Century* (London: Macmillan, 1991).

David Gooding, *Experiment and the Making of Meaning* (Dordrecht and Boston: Kluwer Academic, 1990); discusses Faraday's experimental methods.

The best of the nineteenth century biographies are:

John Tyndall, *Faraday as a Discoverer* (London, 1868).

Sylvanus P. Thompson, *Michael Faraday, his Life and Work* (London, 1898).

On the Royal Institution see:

Morris Berman, *Social Change and Scientific Organisation: The Royal Institution, 1799–1844* (London: Heinemann, 1978).

Gwendy Caroe, *The Royal Institution, an Informal History* (London: John Murray, 1985).

For general studies of science in the nineteenth century see:

J. T. Merz, *A History of European Thought in the Nineteenth Century* (4 vols, Edinburgh, 1904–12; reprinted New York: Dover, 1965), especially volumes 1 and 2.

David Knight, *The Age of Science: The Scientific World-View in the Nineteenth Century* (Oxford: Blackwell, 1986).

Several articles in R.C. Olby, G.N. Cantor, J.R.R. Christie and M.J.S. Hodge (eds), *Companion to the History of Modern Science* (London and New York: Routledge, 1990).

Other relevant works:

Sophie Forgan (ed), *Science and the Sons of Genius: Studies on Humphry Davy* (London: Science Reviews, 1980).

Brian Bowers, *A History of Electric Light and Power* (Stevenage: Peter Peregrinus, 1980).

Crosbie Smith and M. Norton Wise, *Energy and Empire: A Biographical Study of Lord Kelvin* (Cambridge: Cambridge University Press, 1989).

Martin Goldman, *The Demon in the Aether: The Story of James Clerk Maxwell* (Edinburgh: Paul Harris, 1983).

A. Pais, '*Subtle is the Lord . . .': The Science and the Life of Albert Einstein* (Oxford: Oxford University Press, 1982).

Index